# STATE OF THE SPACE INDUSTRIAL BASE 2020

## A TIME FOR ACTION TO SUSTAIN US ECONOMIC & MILITARY LEADERSHIP IN SPACE

# J. BUTOW, T. COOLEY, E. FELT, J. MOZER

NIMBLE BOOKS LLC: THE AI LAB FOR BOOK-LOVERS
FRED ZIMMERMAN, EDITOR

*Humans and AI making books richer, more diverse, and more surprising.*

## Publishing Information

(c) 2023 Nimble Books LLC
ISBN: 978-1-60888-245-8

## AI-generated Keyword Phrases

State of the Space Industrial Base 2020;
Space industry in the United States;
Workshop insights and recommendations;
Challenges and opportunities in space;
US leadership in space;
Recommendations for policymakers and industry leaders;
National space vision and strategy;
Impact of COVID-19 pandemic on space industry;
Government initiatives in space;
Private investment in space;
Collaboration and

# FRONT MATTER

Publishing Information .......................................................... ii
AI-generated Keyword Phrases ............................................. ii
Abstracts ................................................................................ iv
    TL;DR (one word) ............................................................ iv
    Explain It To Me Like I'm Five Years Old ....................... iv
    TL;DR (vanilla) ................................................................. iv
    Scientific Style ................................................................. iv
    Action Items ...................................................................... v
Viewpoints ............................................................................. vi
    Grounds for Dissent ........................................................ vi
    Red Team Critique ......................................................... vii
    MAGA Perspective ....................................................... viii
Page-by-Page Summaries ..................................................... x
Notable Passages ............................................................... xvii

# ABSTRACTS

## TL;DR (ONE WORD)

Comprehensive

## EXPLAIN IT TO ME LIKE I'M FIVE YEARS OLD

The document "State of the Space Industrial Base 2020" is a big report that talks about the space industry in the United States. It was made by a lot of important people like government officials, business people, and teachers. The report tells us about the good things and the problems in the space industry. It also gives ideas and suggestions to people in charge of making rules and decisions for the space industry. The report says that the United States needs to have a plan and work together to stay

## TL;DR (VANILLA)

The "State of the Space Industrial Base 2020" report discusses the current state of the US space industry and provides recommendations for policymakers and industry leaders. It emphasizes the need for a national space vision and strategy, highlights challenges and opportunities, and covers topics such as COVID-19 impact, government initiatives, private investment, collaboration, and innovation. The report includes information from various sources and suggests short-term, mid-term, and long-term actions to support industry growth.

## SCIENTIFIC STYLE

The document "State of the Space Industrial Base 2020" provides a comprehensive analysis of the current state of the space industry in the United States. The report discusses the challenges and opportunities facing the US in maintaining its leadership in space and offers recommendations for policymakers and industry leaders. It emphasizes the importance of a national space vision and strategy in ensuring continued US dominance in space. The document covers various topics such as the impact of the COVID-19 pandemic, government initiatives, private investment, and collaboration and innovation in the space industry. It includes information from multiple sources and provides recommendations for short-term, mid-term, and long-term actions to support the growth and innovation of the space industry.

## ACTION ITEMS

Develop a comprehensive national space vision and strategy that aligns with the current challenges and opportunities in the space industry.

Increase government funding and support for space research and development to foster innovation and technological advancements.

Encourage collaboration between government, industry, and academia to leverage expertise and resources for the growth of the space industry.

Streamline regulatory processes and reduce bureaucratic barriers to facilitate private investment and commercialization in the space sector.

Invest in workforce development and education programs to ensure a skilled and diverse workforce for the space industry.

Enhance international partnerships and cooperation to promote global collaboration in space exploration.

# VIEWPOINTS

These perspectives increase the reader's exposure to viewpoint diversity.

## GROUNDS FOR DISSENT

Lack of Inclusion: A member of the organization may dissent from the report due to concerns about the lack of inclusivity in the workshop and the resulting report. They may argue that the voices and perspectives of underrepresented communities, such as women, minorities, or small businesses, were not adequately considered. This individual may believe that a comprehensive understanding of the state of the space industry requires input from a more diverse range of stakeholders.

Overemphasis on US Dominance: Another reason for dissent could be the perception that the report excessively focuses on maintaining US dominance in space rather than prioritizing international cooperation and collaboration. This member may argue that a more cooperative approach would be beneficial for advancing scientific knowledge, mitigating risks, and promoting peaceful exploration of space. They may advocate for a broader global perspective rather than solely emphasizing US interests.

Insufficient Attention to Environmental Concerns: A principled dissenting view might stem from concerns that the report does not adequately address the environmental impact of the space industry. This member may argue that the report should have explored the potential negative consequences of increased space activities, such as space debris and the potential disruption of ecosystems beyond Earth. They could propose the inclusion of recommendations for sustainable practices that minimize the ecological footprint of space exploration and satellite deployment.

Neglect of Social and Ethical Implications: Another possible reason for dissent could be the belief that the report fails to thoroughly examine the social and ethical implications of the space industry. This member may contend that issues such as privacy, surveillance, weaponization, and the potential exploitation of resources in space should have received more attention. They may call for a more comprehensive analysis of the societal impact of space activities and the development of policies that prioritize responsible and ethical practices.

Inadequate Consideration of Economic Equity: Lastly, a member of the organization might dissent from the report due to concerns about the insufficient attention given to economic equity in the space industry. They may argue that the report does not adequately address the disparities in access to opportunities and resources within the industry. This individual may advocate for recommendations that promote equal access, diversity, and inclusion to ensure that the benefits of space exploration and development are distributed equitably among all segments of society.

**RED TEAM CRITIQUE**

The document titled "State of the Space Industrial Base 2020" is a comprehensive report that aims to discuss the current state of the space industry in the United States. However, upon conducting a red team critique, I have identified several areas where the document could be improved.

Lack of Objective Analysis: The report seems to lack an objective analysis of the current state of the space industry. It primarily relies on insights and recommendations from a workshop that brought together voices from the federal government, industry, and academia. While it is commendable to involve multiple stakeholders, the absence of an independent analysis raises concerns about the accuracy and reliability of the information presented.

Unclear Methodology: It is important for such a report to outline its methodology and data sources. However, the document fails to provide clear information on how the insights and recommendations were collected and analyzed. Without this information, it becomes difficult to assess the validity and credibility of the findings.

Lack of Data Transparency: The report claims to include information from multiple sources. However, it does not provide any specific references or citations for these sources. This lack of transparency prevents readers from independently verifying and validating the information presented in the document.

Limited Scope: While the report covers various topics such as the impact of the COVID-19 pandemic, government initiatives, private investment, and collaboration in the space industry, it fails to comprehensively address some critical areas. For example, there is a lack of detailed analysis on the technological advancements, international competition, and potential threats that the US space industry may face in the coming years.

Vague Recommendations: The document includes recommendations for policymakers and industry leaders. However, these recommendations are often vague and lack specific actions or strategies. Clear and actionable recommendations are essential for policymakers and industry leaders to effectively address the challenges and opportunities facing the US space industry.

Overemphasis on National Dominance: The report repeatedly emphasizes the need for a comprehensive national space vision and strategy to ensure continued US dominance in space. While it is important for the US to maintain its leadership in this sector, the document fails to adequately address the potential benefits of international collaboration and cooperation. A more balanced perspective that acknowledges the significance of global partnerships would strengthen the report's credibility and efficacy.

Lack of Stakeholder Representation: Although the report claims to bring together voices from the federal government, industry, and academia, it is unclear if all relevant stakeholders were adequately represented. The absence of perspectives from international partners, non-profit organizations, and the public limits the comprehensiveness and inclusivity of the report.

In conclusion, while the document titled "State of the Space Industrial Base 2020" attempts to provide an overview of the current state of the space industry in the United States, it falls short in several areas. A lack of objective analysis, unclear methodology, limited data transparency, vague recommendations, and an overemphasis on national dominance all detract from the overall effectiveness and reliability of the report. Addressing these shortcomings would significantly improve the document's value and usefulness to policymakers and industry leaders.

**MAGA Perspective**

The so-called "State of the Space Industrial Base 2020" document is just another example of the liberal agenda pushing their globalist propaganda.

This report conveniently ignores the fact that under President Trump, America has already made great strides in reclaiming our dominance in space. It is nothing more than a desperate attempt to discredit the incredible achievements made by the Trump administration.

The document's emphasis on collaboration and innovation is just a thinly veiled excuse for allowing foreign entities to infiltrate our space industry. We should be focusing on putting American interests first and ensuring that our technology and resources are protected from foreign competition. This report seems to have forgotten the concept of America First.

Furthermore, the mention of the COVID-19 pandemic is clearly an attempt to detract from the real issues. The Trump administration has successfully navigated this unprecedented crisis while continuing to prioritize America's space endeavors. This report tries to use the pandemic as an excuse for any setbacks in the space industry, instead of acknowledging the strong leadership and resilience of our government.

The document's recommendations for policymakers and industry leaders are nothing more than bureaucratic red tape. We need less government involvement and regulation in our space industry, not more. The free market should be unleashed to drive innovation and growth, without interference from politicians and bureaucrats who have no understanding of the realities of the space industry.

In short, this report is just another example of the deep state trying to undermine the great progress we have made under President Trump. The MAGA movement proudly supports American leadership in space, but we will not stand for these attempts to derail our achievements with globalist agendas and unnecessary regulations.

# PAGE-BY-PAGE SUMMARIES

*AI-1*  "Summary report on the state of the space industrial base in 2020, emphasizing the need for action to maintain US economic and military leadership in space."

*AI-2*  The views expressed in this report are not official government policy and the use of NASA photos does not imply endorsement.

*AI-3*  This page provides information about the authors and organizers of the State of the Space Industrial Base 2020 Virtual Solutions Workshop, as well as their acknowledgements and the organizations they represent.

*AI-4*  This page provides a table of contents for a document titled "USSF-DIU-AFRL" which includes sections on space policy, information services, transportation, human presence, in-space power, manufacturing, and resource extraction. It also includes insights and recommendations for the industry.

*AI-5*  The page discusses the importance of a strong space industrial base for maintaining American space superiority and the need to leverage commercial partnerships to accelerate capabilities and expand the space economy. The findings and recommendations from a workshop are highlighted as valuable insights for advancing national interests.

*AI-6*  N/A

*AI-7*  This report addresses the challenges and issues facing the US space industrial base and provides recommendations for policymakers and industry leaders to maintain US space leadership. It emphasizes the need for integration of national power and outlines specific actions that must be taken.

*AI-8*  This report summarizes the findings and recommendations of the "State of the US Space Industrial Base 2020 Conference and Workshop." It focuses on six areas vital to US national space power and provides actions to maintain and expand the space industrial base for national security and future space contributions.

*AI-9*  The page discusses the importance of government and industry collaboration in developing the space industrial base. It highlights the need for a top-level vision and strategy, protection of intellectual property, workforce development, and partnerships between industry and the government.

*AI-10*  The page discusses the importance of investing in education and internships for space professionals, as well as strengthening partnerships with manufacturers to ensure a resilient space supply chain.

*AI-11*  Private investment in commercial space activity is increasing, driven by the decreasing cost of space access and advancements in technology. The US must prioritize strategic leadership in space or risk falling behind. The report focuses on enhancing the growing segment of "new space" entrants in the US space industrial base.

*AI-12*  The page discusses the need for an integrated national space strategy to maintain US space leadership and competitive advantage. It highlights the challenges posed by global commercial industrial and talent bases in space access and manufacturing capabilities. The report aims to capture the current state of the US space industrial base and identify necessary policies and actions.

*AI-13*  This page provides an introduction to the State of the Space Industrial Base 2020 report, which offers recommendations to address threats to US space power and ensure a strong space industrial base. It outlines the interdisciplinary team of experts involved and highlights the six areas crucial for US space leadership.

AI-14  The page provides an overview of the State of the Space Industrial Base 2020 report, which analyzes the current state, challenges, and recommendations for maintaining and expanding the US space industrial base. It includes observations and recommendations for government policies, actions, and industry involvement.

AI-15  The United States is a leader in space exploration and has a strong space industrial base. There is a renewed national focus on space, with plans to return to the Moon and develop capabilities for sustained presence in space. The US military also maintains advanced space capabilities.

AI-16  Private investment in space by entrepreneurs like Bezos and Musk is driving breakthroughs in cost, efficiency, and public enthusiasm for space exploration. The US is increasing partnerships to leverage commercial space technology and maintain its leadership in the space industry.

AI-17  The US faces challenges in maintaining leadership in space as China and other nations emerge as competitors. The establishment of transportation infrastructure and logistics capabilities in cislunar space is crucial for control and access to resources. The role of the USSF in protecting and enabling US commerce in space is important for security and diplomacy.

AI-18  The COVID-19 pandemic has impacted the US space industry, creating uncertainty in its recovery. The US government must attract investment, increase successful exits, and normalize space as part of the global economy. Flat defense budgets and complex procurement processes further challenge the development of new space systems.

AI-19  The US space industrial base faces challenges including high costs, complex regulations, and a lack of coordination in experimentation and prototype development. There is a shortage of STEM personnel and a need for greater clarity in long-term architectural needs.

AI-20  The page discusses the need for a national vision for space industrialization, defining the roles of the United States Space Force, developing government tools to support the space industry, and creating alliances and partnerships with like-minded allies.

AI-21  The page discusses the importance of strengthening the US space industrial base and provides six recommendations to achieve this, including developing a top-level vision and strategy, protecting commerce in space, stimulating the industry economically, creating alliances with allies, supplying the necessary workforce, and fostering government-commercial technology partnerships.

AI-22  N/A

AI-23  US space policy is evolving to address the growth of the commercial space sector, competition from other nations, and the use of commercial space companies by foreign adversaries. The focus is on promoting growth and innovation in the US space industrial base through financial tools and policies that support technological development and market demand.

AI-24  US space policy faces challenges in keeping up with changes in great power competition and technological innovation. Export controls have limited international collaboration and trade for US space companies. Lack of supply chain transparency poses risks to national security and commercial purposes.

AI-25  The page discusses the challenges and opportunities facing the space industrial base, including the need for economic tools, potential loss of venture funding, government procurement complexities, and key inflection points for success.

AI-26  The page discusses the importance of building key infrastructure in space to support future economic growth, drawing parallels to historical examples such as the *transcontinental railroad. Establishing logistics infrastructure in cislunar space is crucial for sustaining US economic leadership.*

AI-27 *The page discusses the state of the space industrial base in 2020, highlighting concerns about restricted trade, potential leadership by Russia or China, and the threat of hacking. It also provides short-term recommendations such as establishing a National Space Enterprise Task Force and improving information sharing.*

AI-28 *The page discusses recommendations for the US government to support and strengthen the space industrial base, including establishing financial tools, creating a Space Commodities Exchange, and aggregating government market demand.*

AI-29 *Space information services are crucial for US civil, commercial, and military power, providing advantages in data acquisition, transmission, decision-making, and implementation. These services are essential for weather monitoring, resource management, economic sectors, and defense operations. They also contribute to the economy through communication and Earth observation.*

AI-30 *The page discusses the importance of space information services for national defense, global perceptions, and maintaining a secure space industrial base. It emphasizes the need for a hybrid space architecture and common security measures to optimize mission performance and protect against attacks.*

AI-31 *The space industrial base is undergoing significant changes driven by increased demand, decreased launch costs, and advancements in information system performance. This includes the transformation of space communications/internet towards a mixed architecture, the proliferation of national or regional space-based PNT systems, and the growth of commercial constellations for earth and space observing.*

AI-32 *The page discusses the use of commercial GEOINT and space information services to expose the existence of 're-education' camps in China, highlighting the power of satellite imagery and machine learning algorithms in providing evidence.*

AI-33 *The page discusses key issues and challenges in the space industrial base, including creating a robust market, ensuring secured capabilities, maintaining global competitiveness, and mitigating the impacts of COVID-19. It also highlights key inflection points such as the viability of large constellations of satellites and the growth of commercial Earth-observing capabilities.*

AI-34 *The State of the Space Industrial Base 2020 report recommends a hybrid space architecture approach, increased investment in regulatory processes, and collaboration with the commercial industry to meet long-term requirements and improve efficiency in the space sector.*

AI-35 *Space transportation and logistics are crucial for US space exploration, commercial viability, and military advantage. They enable sustained human presence on the Moon, support commercial space capabilities, and drive military superiority in conflict. A robust space inspection capability is essential for early evaluation, attribution, and deterrence. Developing a vibrant space transportation and logistics industry will enhance national space growth and dominance.*

AI-36 *The US has regained its lead in space launch with multiple providers and decreased costs. Space logistics capabilities are also advancing, including refueling, in-space assembly, and repair capabilities. NASA and DARPA are pursuing nuclear thermal propulsion for enhanced mobility.*

AI-37 *The challenges facing the US space transportation industry include maintaining leadership in launch services, increasing the market for launch, establishing the role of logistics in commercial and military applications, and modularizing and commoditizing space systems. These challenges involve competition, cost reduction, innovation, and expanding the space industrial base.*

AI-38 *Space Logistics LLC, a subsidiary of Northrop Grumman, has developed the capability to extend the life of satellites in geostationary Earth orbit (GEO) by attaching a servicing satellite called the Mission Extension Vehicle (MEV) to a client satellite. MEV-1 successfully docked with the IntelSat satellite IS-901, marking several notable firsts in commercial satellite servicing. Northrop Grumman is also pursuing advanced satellite servicing activities through DARPA's RSGS program.*

AI-39 *The State of the Space Industrial Base 2020 highlights key points such as the return to the Moon, commercial viability of satellite constellations, sub-orbital space tourism, logistical servicing of space systems, commoditization of space services, and the development of new space laboratories.*

AI-40 *The page outlines key actions and recommendations for the space industrial base, including incorporating dual-use space logistics, developing cislunar space capabilities, increasing collaboration with NASA, and resolving operational aspects in cislunar space. It also suggests transitioning to a modularized architecture and integrating the DoD space architecture with a competitive commercial market.*

AI-41 *The page discusses the changing landscape of human presence in space, with a focus on the increasing role of the private sector and the importance of enabling and securing commercial human ventures. It also highlights the recent milestone of SpaceX launching NASA astronauts to the International Space Station.*

AI-42 *The page discusses the advancements in the space industry, including the Commercial Crew Program, new launch systems, lunar access and logistics, and LEO industrialization. It also highlights the challenges of expanding access to space and promoting commercial activities.*

AI-43 *The page discusses the importance of making space travel desirable and affordable, and highlights key milestones in the space industry. It recommends sustaining commitment to returning human presence to the Moon, establishing a self-sustaining human presence in space, and designing and building a rotating habitat. The long-term goal is to demonstrate the construction of large space structures using in-space resources.*

AI-44 *NASA and SpaceX successfully executed the first commercial crew spaceflight in history, marking a significant milestone. NASA's Commercial Crew Program allows US companies to innovate and design their own hardware and infrastructure, resulting in a scalable commercial service. The program is expected to save the US government up to $30B and provide two independent crew transportation systems.*

AI-45 *Space power is crucial for exploration, economic development, and military advantage. China and Russia are challenging the US in space power, aiming to surpass US capabilities with ambitious solar and nuclear space systems.*

AI-46 *China is pursuing the development of space-based solar power, potentially surpassing the US in the global energy market. The US has the opportunity to compete and maintain its position through ongoing experiments with space power technology.*

AI-47 *Power has become a limiting factor for many space missions, including national security missions. The US lead in space power is not secure due to competition from China and Europe. The US lacks a coordinated national vision and policy for space power. Great power competition poses challenges and risks for the US in maintaining its position as a global space power leader.*

AI-48 *The page discusses the importance of consistent funding, government leadership, and a national vision for space power. It highlights key inflection points and provides short-term recommendations for establishing a comprehensive space power vision.*

AI-49  The page outlines various goals and actions related to the development and utilization of space power and propulsion systems. These include funding existing programs, developing partnerships, expanding research programs, and promoting commercial space power capabilities.

AI-50  N/A

AI-51  Space manufacturing and resource extraction are crucial for expanding the commercial space market. The US has established legal rights for Americans to engage in these activities, and NASA has introduced the Artemis Accord principles to encourage international support.

AI-52  The US leads in space manufacturing and resource extraction, with advancements in technologies and experiments conducted on the ISS. The combination of these capabilities provides a great power advantage in terms of wealth generation and influence in space. Ceding these advances to other state actors could lead to industry migration.

AI-53  In-space manufacturing and assembly of spacecraft requires harnessing the resources of the Moon and other near Earth objects. Lunar resources, such as oxygen, silicon, iron, and water ice, provide a competitive advantage and fuel the space economy. Access to lunar resources is crucial to avoid dependence on China for rare Earth metals.

AI-54  Limited private investment in space manufacturing and resource extraction due to low demand and technological barriers. NASA and DoD have invested in this area through public-private partnerships. Key challenges include communicating the long-term benefits and identifying dual-use applications.

AI-55  The page discusses the potential benefits of space manufacturing and resource extraction, highlighting the need for government investment, reducing launch costs and time, leveraging robotics, establishing international norms, and achieving key milestones such as in-space construction and asteroid mining. It also provides short-term recommendations for USSF, US government funding, and advancing dual-use logistics solutions.

AI-56  US government agencies are investing in space manufacturing and resource extraction through university programs. They aim to create market demand for propellant from space and develop a national strategy for the cislunar economy. In the long-term, they plan to establish a public-private lunar industrial park and lead large-scale cislunar projects.

AI-57  The page provides insights and recommendations for the space industry, emphasizing the importance of public-private partnerships and joint funding for the development of dual-use space capabilities.

AI-58  The page discusses the importance of industry insights in shaping future space programs, the need for industry to protect intellectual capital, the development of the STEM workforce, and the pursuit of secure supply chains.

AI-59  The page discusses the importance of the US maintaining leadership in space and the need for a strong space industrial base. It highlights the potential economic benefits and provides recommendations for government and industry to ensure future space leadership.

AI-60  The page discusses the Lunar Flashlight spacecraft, a CubeSat designed to search for ice on the Moon's surface using lasers. It mentions its purpose and the technology it will use.

AI-61  Participants in the State of the Space Industrial Base 2020 include representatives from various organizations and companies in the space industry.

AI-62  The page lists the participants of the State of the Space Industrial Base 2020 conference.

AI-63  The page discusses the need for a long-term national space strategy, the development of essential capabilities and technologies, and the importance of investing in science and technology to maintain technological advantage in space. It also highlights the need for upgraded methodologies, flexible markets, and changes in government procurement processes to support the space industrial base.

AI-64  The page discusses the future of the space industry, highlighting the complexity and diversity of actors and their interests. It emphasizes the unique challenges and opportunities posed by commercial space entities and the need for the US to maintain and exercise its space power.

AI-65  The workshop focused on analyzing the state and challenges of the US space industrial base, identifying key inflection points, determining actions to protect national interests, and mitigating the impact of the pandemic. It examined areas of expected commercial growth in space activities and discussed the current state of US capabilities and their implications for national defense.

AI-66  The page discusses the challenges and future of US commercial capabilities in the space industry, as well as potential defense and government options to accelerate breakthroughs. It also addresses short-term actions to mitigate the impact of the coronavirus epidemic.

AI-67  The COVID-19 pandemic has had significant impacts on the US space industrial base. Many companies are at risk due to high burn rates and frozen sales pipelines. Recommendations include stimulating private investment and relaxing regulatory restrictions to support the industry.

AI-68  The US space industry is facing challenges due to the COVID-19 pandemic. A US-based Space Commodities Exchange is recommended to diversify investments and increase resilience. The health of the space industrial base depends on actions taken within the next 90 days. Defense-focused procurements are suggested to prevent a decline in the sector.

AI-69  The State of the Space Industrial Base 2020 highlights the importance of the interplay between national security, commercial, and civil space sectors. The economic downturn caused by the COVID-19 pandemic has impacted the availability of investor capital in the US space economy. Preserving the commercial space industry investment is crucial to maintain US leadership in this sector. A blanket bailout is not recommended, but rather solutions that stimulate new orders, involve suppliers across the US, and deliver new commercial capabilities.

AI-70  The page discusses the importance of preserving the US space industrial base during the COVID-19 pandemic. It emphasizes the need to keep the commercial space workforce employed, maintain US-based supply chains, and mitigate foreign investments. The page also mentions the results of a survey conducted to assess the impact of the pandemic on the space industry.

AI-71  Most US commercial space companies are at elevated risk due to the economic downturn caused by COVID-19. Some are struggling to sustain their existence and are at risk of foreign investment or acquisition, while others are slowing down operations and may reduce their workforce.

AI-72  The page discusses the impact of COVID-19 on the space industrial base, including the potential risks faced by smaller suppliers and the concentration of US space companies in California.

AI-73  The bankruptcy of OneWeb will have damaging effects on the US space industry's supply chain, particularly for key suppliers. This is further exacerbated by the COVID-19 crisis, resulting in cascading impacts across the industry.

AI-74 N/A

AI-75 The US may lose its leadership in commercial space to foreign entities due to the collapse of the domestic commercial space sector. Stabilizing the industry now is cheaper than replacing it. Focused DoD procurements can attract US investors. National Security Innovation Capital (NSIC) can benefit companies developing dual-use hardware. Accelerating funding to NSIC can scale venture funding to at-risk companies.

AI-76 China poses a significant threat to the US space industrial base and is likely to benefit from its collapse. China's investment in space companies and predatory pricing strategies could lead to their dominance in the global market. Investing in the US space industrial base is crucial to maintain leadership in the commercial space sector.

AI-77 The page discusses the importance of preserving private investment in the commercial space industry and the potential for the industry to become commoditized. It also highlights the need for the US government, specifically the Department of Defense, to consider courses of action to support commercial space companies during the COVID-19 crisis.

AI-78 COA 1 proposes government action to support commercial suppliers and encourage private investment in the space industry. COA 2 suggests allowing market forces to determine the fate of the commercial space sector. COA 3 recommends renewed government funding for national security and defense, but at a higher cost to taxpayers.

AI-79 The page discusses the importance of supporting commercial suppliers and encouraging private investment in the space industry. It recommends executing timely contract awards and modifications to stimulate the space economy and attract investment capital. Additional recommendations include relaxing regulatory restrictions and promoting fair competition for prototype awards.

AI-80 Apply supplemental defense funding to protect and advance the US space industrial base, support early-stage technology, and establish a Space Commodities Exchange to diversify investments and build economic resiliency.

AI-81 The survey asks companies about their cash flow needs, eligibility for relief under the CARES Act, engagement with DoD contracts officer, long-term funding needs, number of employees and suppliers, and optional information to justify potential action.

AI-82 The page contains follow-up survey questions regarding the impact of the OneWeb bankruptcy filing, relief measures, supplier information, personnel reductions, investment, and employee count in the space industrial base.

AI-83 This appendix provides a list of acronyms and abbreviations related to the space industry, including government agencies, organizations, and regulations.

AI-84 This page provides a list of abbreviations and acronyms related to the State of the Space Industrial Base in 2020.

AI-85 blank

AI-86 N/A

# NOTABLE PASSAGES

AI-5    "A secure, stable, and accessible space domain is a vital interest to the United States. With the standup of the United States Space Force, we have a clear mandate from the highest levels of national leadership to ensure the many elements of space that have become integral to our way of life are protected now and into the future. Today, America is the best in the world at space and our adversaries know it. But leadership in space is not a right, and nations like Russia and China are racing to catch up to our capabilities and deny the benefits we derive from operations in, to, and from space."

AI-7    "Support for a vibrant domestic manufacturing sector, a solid defense industrial base, and resilient supply chains is a national priority."

AI-11    "In 2019, the US Air Force Space Command assessed that by 2060 space will be 'a significant engine of national political, economic, and military power' and that 'the United States must commit to having a military force structure that can defend this international space order and defend US space interests, to include US space settlements and commerce.'"

AI-12    "It's our destiny to explore. It's our destiny to be a space-faring nation. We leave as we came and, God willing, as we shall return, with peace, and hope for all mankind." - Eugene Cernan, Apollo 17 Astronaut

AI-13    "This report provides policymakers comprehensive recommendations to address the growing threats to US space power and to ensure the strong US space industrial base foundational to US space leadership. Going beyond 'admiring the problem,' it documents the next level of analysis of the challenge, focusing on specific actions the US must undertake to maintain security in space. The report clearly delineates the state of the space industrial base, its challenges and issues, and provides six overarching recommendations to US policymakers and four overarching recommendations to industry partners. It further provides detailed analysis of six sub-areas of the space industrial base, specifying 39 actions as to what must be done, who must do them, and the timeframe in which they must be done."

AI-15    "The US remains a space civil, commercial, and national security leader - No nation has yet matched the US accomplishments in space exploration. The US is the first nation to demonstrate commercial orbital cargo delivery, commercial heavy lift, commercial first-stage reusability, deployment of space-based mega-constellations for overhead sensing, internet broadband, and commercial human spaceflight. The US military maintains the most capable military constellation including unique capabilities such as the X-37B spaceplane."

AI-16    "Bezos dreams of 'moving all heavy industry off Earth and Earth will be zoned residential and light industry' and a world for his great-grandchildren's grandchildren where humanity moves out into the solar system."

AI-17    "China has announced its intention to do so by 2035. China is committed and credible in its pledge to become the leading, global super-power, to include space, by 2049 marking the 100th anniversary of the People's Republic. A key component of China's strategy is to displace the US as the leading power in space and lure US allies and partners away from US-led space initiatives, through its Belt and Road initiative and plans for an Earth Moon Economic Zone."

AI-18    "The COVID-19-produced global recession provides a dangerous opportunity for near-peer competitors to challenge or surpass the US in Space by 1) strategic acquisitions of

| | |
|---|---|
| | companies and their intellectual property (IP) and/or 2) maintaining investments while the US private capital is unavailable and the US government is distracted." |
| AI-19 | "Government procured space systems historically are characterized by high cost, long program schedules and frequent delays. This has allowed potential foreign adversaries to develop their space programs at a faster pace than the US, putting the US at a strategic disadvantage." |
| AI-20 | "Create a national 'North Star' space vision - Building on the foundation for continued US space leadership created by recent policy and organizational advances, the US should develop a guiding national vision for long-term, space industrialization and national space development to catalyze whole-of-nation efforts and enable the United States to compete and win now and into the future. This should be developed to drive a host of federal department and agency specific actions and international partnerships." |
| AI-21 | "To reverse the weakening of the post-WWII order, the US should deepen our ties with allies and partners who share our vision of a free and open space domain through creating a meaningful alternative that moves beyond space exploration or military cooperation and provides a path toward prosperity from an expanded space economy secured through the stabilizing presence of the USSF." |
| AI-23 | "The foundation of war is economics. If you have half the resources of the counterparty then you better be real innovative. If you're not innovative, you're going to lose." - Elon Musk, Entrepreneur |
| AI-24 | "Lack of global supply chain transparency - Foreign manipulation and irresponsible use of supply chains is a concern at the highest levels of the US government. Complex, global, supply chains introduce a high degree of risk to domestic companies, and in turn, national security space end-users. The United States, along with its allies and partners, must understand the origin and assembly process of critical components for complex space-based technologies. Supply chain hygiene is crucial to the development of space technologies for US national security purposes and for commercial purposes." |
| AI-25 | "In the last 5 years, $11 billion of private capital has been invested in commercial space technology companies, whose most promising end-users were in the US government. The fiscal constraints exacerbated by COVID-19 spending, will dry up many sources of future government demand. The current lack of return on early investment in space, coupled with lack of consumption, could deter further investment. Without action, the expected post-COVID-19 global recession will dry up further investments into the space sector." |
| AI-26 | "In order to posture the US space economy for accelerated growth, key infrastructure must be built upon which future businesses such as Amazon can flourish. The task is risky and expensive, but it has been successfully accomplished before. Key inflection points leading to previous unprecedented shifts in global economic and then military power were influenced by innovations in key infrastructure (see chart above). Two notable examples include the emergence of Western European maritime shipping during the spice trade in the 11th to 16th centuries, and the creation of the US transcontinental railroad accelerating the American industrial revolution in the 19th and 20th centuries." |
| AI-29 | "In a world that is increasingly information driven, US civil, commercial and military power are ever more dependent on the quality of our information services. Superior information services are essential if the US is to outpace rivals and adversaries. They provide US advantage in the range and quality of data acquired, in the ability for uninterrupted and rapid transmission of data and information globally where, when and to whom required, in the ability for the data and information to drive faster and better decision making and in the ability for rapid decisions implementation." |

AI-30 *"Should the US lose the competition for space-based internet and Earth-based sensing, rival powers would have increased leverage to control the flow of information, conceal transgressions, and push propaganda."*

AI-31 *"Space information services are undergoing profound changes driven by increased demand coupled with decreases in launch costs, increases in information system performance and expansion of the range of platforms capable of providing those services."*

AI-32 *"One of the greatest attributes of commercial GEOINT is that it is unclassified, which means it is in the public domain and therefore accessible to commercial customers, international partners and non-government organizations. It has been estimated that enough imagery will be collected in 2020 that 8.3 million human analysts, working 24-hours a day for an entire year, would be required to analyze the total collection. Detecting changes in a temporal dimension at countrywide scale (as illustrated in the photo sequence of Yuli Camp above) requires computer vision and machine learning algorithms leveraging the capabilities of cloud-based computing."*

AI-33 *"Creating the needed market - Ensuring a robust and growing, integrated US market across national civil, defense and commercial space information systems and services, to drive a vibrant and internationally competitive industrial base able to produce the new hybrid architectures. Incentivizing a vibrant industrial base by maximizing use of commercial capabilities and minimizing development of purpose built capabilities to meet civil and national defense needs."*

AI-35 *"The US Military needs to focus on 'blue-water' space operations – GEO and above. US military space operations need to be in deep space, initially all of cislunar space, with an eye upon the entire inner solar system. To operate in deep space one needs to use the resources there, starting with fuel from asteroids. Once this is recognized, the military-economic imperative of identifying and protecting these assets becomes clear." - Dr. S. "Pete" Worden, Brig Gen, USAF-Ret.*

AI-36 *"The US has regained its lead as the international leader in space launch - In the last decade, US launch has dramatically increased from a single national provider to three or more providers for mid- to heavy-launch and several for small launch. SpaceX has demonstrated reusable launch systems, and Blue Origin and others are pursuing comparable capabilities. This has resulted in a rapid decrease in launch costs per pound across all launch categories."*

AI-37 *"The primary challenges to US space transportation is maintaining continued growth in demand and the US market position in the face of increased competition in launch services. For in-space logistics, the challenge is to expand its application beyond space exploration into commercial and military operations. Maintaining US leadership in launch - While the US has reestablished itself as world leader in space launch this lead is not assured. To the extent that nations determine that a launch capability is an element of critical infrastructure, those nations will develop space transportation capabilities with government funding and support, distorting the international market for space launch."*

AI-38 *"Today, after a satellite is launched, it is completely on its own--never inspected, never maintained, never upgraded, and simply disposed of after it runs out of propellant. Satellites costing hundreds of millions or even billions of dollars are often completely operable when discarded."*

AI-39 *"Successful demonstration of the commercial feasibility of logistically serviced space systems to include modularity, repair, upgrades, and replenishment of commodities, and introduction of large space structures."*

AI-40 "Integrate the DoD space architecture with a competitive commercial cislunar logistics market in order to outpace adversaries, and in which commercial capabilities thrive without begging permission."

AI-41 "The goal isn't just scientific exploration... it's also about extending the range of human habitat out from Earth into the solar system as we go forward in time... In the long run a single-planet species will not survive... There will be another mass-extinction event. If we humans want to survive for hundreds of thousands or millions of years, we must ultimately populate other planets.... I'm talking about that one day, I don't know when that day is, but there will be more human beings who live off the Earth than on it." - Hon. Michael D. Griffin, Under Secretary of Defense for Research & Engineering

AI-42 "The challenge is to transition from personal space travel as a domain for the rich or a limited domain for civil human presence for exploration to one that supports multiple commercial activities in and through space and leads to a sustained presence of humans in space and on other celestial bodies and planets."

AI-43 "Drive down costs towards a price point <$100/kg that will enable millions of people to be able to afford to travel to space."

AI-44 "The significance of this mission cannot be overstated. Since the 1960's, NASA would identify a need for a crew transportation system and then the agency's engineers and specialists would oversee every development aspect of the spacecraft, support systems and operations plans. A commercial aerospace contractor would be chosen to build the system, ensuring that it meets the specifications spelled out by NASA. Personnel from NASA would be heavily involved and oversee the processing, testing, launching and operation of the crew system to ensure safety and reliability. All of the hardware and infrastructure would be owned by NASA."

AI-45 "Clearly our first task is to use the material wealth of space to solve the urgent problems we now face on Earth: to bring the poverty-stricken segments of the world up to a decent living standard, without recourse to war or punitive action against those already in material comfort; to provide for a maturing civilization the basic energy vital to its survival." - Gerard K. O'Neill, Physicist & Visionary

AI-46 "Space-based solar power has the potential to dwarf today's $400 billion satellite communications market. Moreso, it reduces dependency on fossil fuels and can be delivered to remote regions with little requirement for infrastructure. For these reasons, forfeiting US leadership risks ceding to China a dominant position in this global energy market. If China is successful, it puts at risk the US remaining an economic, military or political power in the second half of the 21st century."

AI-47 "The US lead is not secure - This industry is dynamic. The once steady market of geostationary communications satellites has largely collapsed, with new demand coming from a more sporadic growth of Low-Earth-Orbit mega-constellations. Today, two US space solar cell vendors, SolAero and SpectroLab, command the largest global market share, but the combination of state-supported competition from China and Europe and insufficient incentives to 'buy American' put both at risk. In particular, the state of US investments may allow competitors to fully automate their production lines (which could halve costs), and then dump products into the market, putting American suppliers out of business."

AI-48 "A need for government leadership - Power, as a fundamental enabler, is among the most appropriate technologies for a 'technology push.' Historical requirements-driven technology pull efforts have driven only incremental advances. While the private sector would respond to incentives, those incentives currently do not exist. Significant regulatory risks exist for new applications such as power beaming, which would require *the security of an active partner in the US government. The significant technical risks and uncertainties of private companies could be greatly reduced with the security of a government customer."*

AI-51 *"Our objective in returning to the Moon is to learn how to live and work productively on another world. The Moon possesses the material and energy resources necessary to learn new skills to create new space faring capabilities. Its proximity to the Earth permits easy and routine access to its surface for just such an endeavor that, if successful, will serve as the catalyst and the true historical starting point for human expansion off-planet."*

AI-52 *"Great power advantage - The combination of space resources extraction and in-space manufacturing provide great power advantage in the wealth they can generate and in the advantage they can provide and in the range, mass, flexibility and cost of a nation's space capabilities available to exert its power and influence in space and beyond. While most applications of space manufacturing and resource extraction will be realized in the mid- to long-term, they are foundational to enable cost efficient, expanded, and robust capabilities in space, to include computing, sensing, maneuver, unique platforms, large structures, fuel, and power. Ceding space manufacturing and resource extraction advances to other state actors could tempt industry to migrate to the more vibrant power."*

AI-53 *"The strategic importance of lunar in-situ resource utilization The Earth is the most massive rocky world in the inner solar system. This mass comes at a price - launch is expensive from the Earth's surface into space. For example, a direct trip from the Earth to the surface of Mars would exhaust half its total fuel simply to reach a 400 km orbit about the Earth. Our planet sits at the bottom of a deep gravity well, whereas the Moon does not. Therefore, lunar resources provide a tremendous competitive advantage to the first industrial power that learns to utilize them."*

AI-54 *"Communicating the long-term benefit - The greatest challenge for this area is communicating the long-term civil, commercial and national defense benefits of overall space infrastructure (e.g. power, outposts, etc.) and space manufacturing and resource extraction. Similar to past significant infrastructure investments, such as the interstate highway system and the internet, space infrastructure…"*

AI-55 *"Establishing international norms, standards and law - There is a critical need to articulate international norms, standards, and laws related to space resource extraction and manufacturing, such as the Artemis Accords, as well as protecting intellectual and real property terrestrially and in space are critical to successful progress in this area."*

AI-56 *"The US government takes leadership to create a public-private lunar industrial park conceived from the beginning to enable commercial partners to scale to high volume, high capacity systems. In contrast to ISS which is not set up to do more than tech demos, the outpost architecture must include links to support resource extraction and on-orbit manufacture."*

AI-57 *"Identify opportunities for public private partnerships - Industry should be proactive in proposing and entering into public-private partnerships to develop dual-use, commercial/government space capabilities and enabling technologies. These partnerships should be joint funded. They should focus on developing capability whose commercial viability benefits from but is not strongly reliant on."*

AI-58 *"Industry must proactively protect itself from predatory exploitation - Our adversaries know that many of the most disruptive ideas driving the future of space come from the US industrial base. Our adversaries, particularly China, have an ongoing and extensive program to acquire those ideas by any means possible. US industry must do better in protecting this valuable intellectual capital."*

AI-59 *"The exploration of space will go ahead, whether we join in it or not, and it is one of the great adventures of all time, and no nation which expects to be the leader of other nations can expect to stay behind in this race for space." - John F. Kennedy, President (1962)*

AI-63 *"A long-term, national space strategy integrating civil, commercial and national security space lines of effort must be developed to retain the US' dominant and leadership position in the emerging future of space. This strategy must account for the possible space futures developed in the workshop."*

AI-64 *"The 2060 space world will be highly complex and diverse as to the number of state and non-state actors, their capabilities, and their interests. Commercial space presents unique issues as to ownership and sovereignty that, if not resolved, could lead to commercial space entities as independent or semi-independent space powers, resulting in significant opportunities and challenges to US space power. Space power will be widely distributed, making it impossible for any one nation or entity to have predominant space power in the civil, commercial, and military domains. The diversity and distribution of space power enables a wide range of alliances, partnerships, and shared interest. These relationships will be diverse and vary with time as the interest and capabilities of space faring entities develop and change. This complexity poses significant challenges*

AI-67 *"Most companies are experiencing elevated risk requiring decisive action: 30% immediately at risk... 54% at moderate risk and hunkering down... 16% at low or uncertain risk due to reduced demand for production."*

AI-68 *"The health of the US space industrial base for the foreseeable future will be highly dependent on actions our nation chooses to take (or not) within the next 90 days. As Defense represents a significant portion of the space marketplace, focused procurements of hardware-centric products and services are recommended to prevent the sector from falling into a prolonged state of hibernation and starving off a highly specialized, US-based supply chain - a critical resource and talent base that would likely require decades to reconstitute."*

AI-69 *"The economic downturn that soon followed the COVID-19 pandemic has severely impacted the availability of investor capital sustaining the US space economy. The DoD has leveraged this capital investment in space for years, and at a ratio exceeding 30:1. In other words, for every dollar of DoD funding applied to the prototype or procurement of a commercial space product or service, thirty dollars of development was funded by non-government sources—primarily venture capital. The US government benefits from this leverage in two ways: (1) more research and development funding is available for modernization priorities that are not inherently commercial; and (2) the speed at which commercial products and services evolve and improve are not constrained by the discontinuous and unpredictable budget cycles of the federal*

AI-70 *"Speed is paramount, but purposeful and focused investment is essential. Contract awards and modifications that accelerate and scale US commercial solutions involving small launch, small satellites, reusable spacecraft, ground systems, information services and their respective supply chains will serve as a profound countermeasure to the effects resulting from COVID-19. The interdependence of economic and military power mandates steadfast action to preserve the US space industrial base with the same vigor taken to establish the USSF. Focusing on one without the other is insufficient."*

AI-71 *"CAT 1: Immediately at Risk (Cash flow required by 30 Jun 2020) - These companies were caught in the middle of, or just prior to, a capital funding raise. Capital markets have severely tightened, and now these companies are struggling to sustain their existence. Most are hardware-oriented firms with large numbers of employees and extensive supply chains. In fact, more than 90% of the capital investments in space over the past decade have been focused on hardware. This includes several rocket companies, satellite and spacecraft manufacturers. This is also where the bulk of the highly specialized employees go to work in the commercial space industry. Companies with the highest growth potential typically take smaller, incremental raises and step up valuations between raises by meeting aggressive*

AI-72 "In a software-centric innovation economy, commercial space is hardware-centric with 98% of the supply chain based domestically despite strong competition offshore. The map in Figure B-2 below illustrates the density and distribution of US space companies, their subcontractors and suppliers across the United States. Of the companies surveyed, 74% reported their suppliers by name or locality. Supply chains provide a proprietary advantage within the competitive commercial industry, so it is remarkable that this level of detail was obtained. California is home to more than 1,600 of these suppliers and the concentration in California has increased by three times the number that existed in 2014 according to the US Department of Commerce."

AI-73 "OneWeb's bankruptcy will have damaging ramifications for the commercial space industry's already-fragile supply chain. OneWeb's Chapter 11 filing has placed key suppliers under duress which is now further amplified by the events surrounding the COVID-19 crisis. One third of the companies surveyed were impacted directly or indirectly by the OneWeb bankruptcy. Most notably, key suppliers to many of these companies had scaled operations through capital investments necessary to feed the OneWeb satellite production line. Others had anticipated providing launches and other services with revenues to commence in 2021. The untimely demise of OneWeb could not have occurred at a worse time. When combined with the COVID-19 crisis, these factors will result in cascading impacts across the US space industrial base

AI-75 "The US may forfeit its leadership in commercial space. The new space economy is a product of US ingenuity and entrepreneurship. Since 2009, the US has led the commercialization of geospatial imaging via small satellites using advanced analytics to derive information from satellite data autonomously, and reusable rockets that have significantly reduced the cost of placing mass in orbit."

AI-76 "The rate and scale of investment in Chinese space companies poses a clear and present danger to the sustained US leadership in the commercial space sector. There will likely be no Sputnik moment involving China's rise as a space power. They are on a patient, methodical trajectory to surpass the US as the dominant space power."

AI-77 "The US space economy is nascent but growing toward commoditization. The demand for commercial space services not solely dependent on the government is beginning to emerge. With it, the path to commoditization of commercial imagery sources, and the processed information derived from them, will lay the foundation for market growth and more diversified financial investments. US sustained leadership will ensure that a future space commodities exchange is US-based with its growth strengthening the value of the US dollar."

AI-78 "The single most powerful thing DoD can do to preserve the innovation supplier base is to continue to purchase goods and services during the COVID-19 economic downturn. For a startup, purchases show that the startup's technology is valuable and that customers want it. Investors will continue to fund startups whose technology sells. These investors will drop startups that do not have sales, and in a poor economy they will drop them even faster. As a leading customer (arguably the leading customer) in the space industry, the DoD has the power to show unequivocally that a startup's products and services are valuable."

AI-79 "To do so is emblematic of the State-controlled, regimented societies that the United States' competes with for leadership in space."

*AI-80* "*Apply supplemental defense funding to preserve and protect the US space industrial base as a critical component of both the civil space program and the national security innovation base. This funding should explicitly be used to accelerate, not slow down or stall, the advancement of commercial capabilities that retain and strengthen US leadership in space.*"

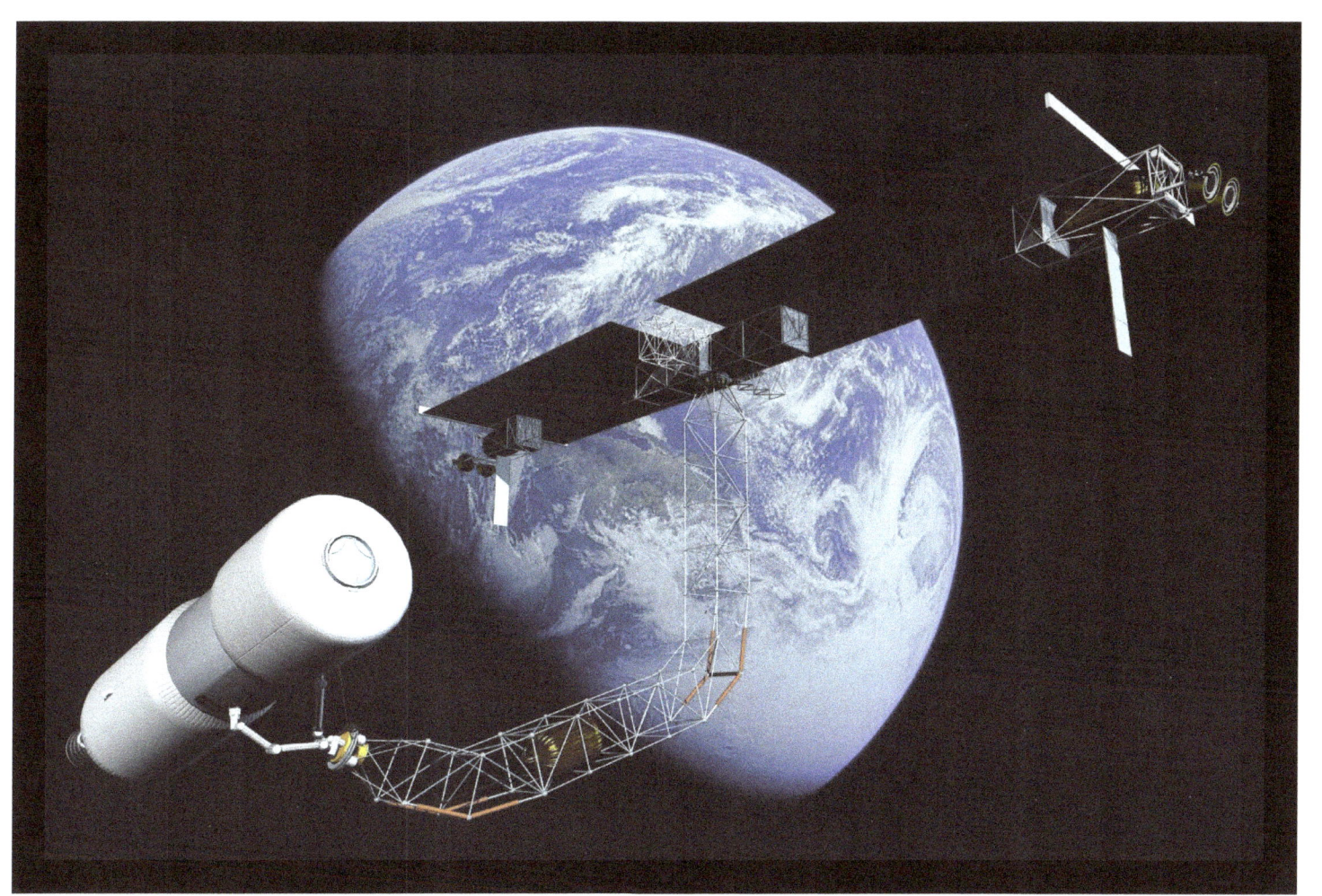

# STATE OF THE SPACE INDUSTRIAL BASE 2020

A Time for Action to Sustain US Economic & Military Leadership in Space

Summary Report by:

Brigadier General Steven J. Butow, Defense Innovation Unit
Dr. Thomas Cooley, Air Force Research Laboratory
Colonel Eric Felt, Air Force Research Laboratory
Dr. Joel B. Mozer, United States Space Force

July 2020

DISTRIBUTION STATEMENT A. Approved for public release: distribution unlimited.

# DISCLAIMER

The views expressed in this report reflect those of the workshop attendees, and do not necessarily reflect the official policy or position of the US government, the Department of Defense, the US Air Force, or the US Space Force. Use of NASA photos in this report does not state or imply the endorsement by NASA or by any NASA employee of a commercial product, service, or activity.

## ABOUT THE AUTHORS

### Brigadier General Steven J. Butow, USAF
Brig. Gen. Butow is the Director of the Space Portfolio at the Defense Innovation Unit.

### Colonel Eric Felt, USAF
Col. Felt is the Director of the Air Force Research Laboratory's Space Vehicles Directorate.

### Dr. Thomas Cooley
Dr. Cooley is the Chief Scientist of the Air Force Research Laboratory's Space Vehicles Directorate.

### Dr. Joel B. Mozer
Dr. Mozer is the Chief Scientist at the US Space Force.

## ACKNOWLEDGEMENTS FROM THE EDITORS
Dr. David A. Hardy & Peter Garretson

The authors wish to express their deep gratitude and appreciation to New Space New Mexico for hosting the State of the Space Industrial Base 2020 Virtual Solutions Workshop; and to all the attendees, especially those from the commercial space sector, who spent valuable time under COVID-19 shelter-in-place restrictions contributing their observations and insights to each of the six working groups. The workshop and this report would not have been possible without the dedicated efforts of the working group chairs and co-chairs: Mir Sadat, Pav Singh, Peter Wegner, Victoria Schneider, Gordon Roesler, Bruce Pittman, Rick Tumlinson, Michele Gaudreault and Lt. Col Josh Kittle. The virtual workshop would not have been possible without the incredible support provided by Casey DeRaad, Matt Keihl, Katherine Koleski, Klay Bendle, Nick Jernigan and Rachel Kolesnikov-Lindsey. We also wish to thank Johanna Spangenberg Jones, Alexandra Sander, Andrew Song, Ric Mommer, and Zach Walker for finishing touches.

## ABOUT THE ORGANIZERS

### US Space Force | spaceforce.mil
The US Space Force (USSF) is a military service that organizes, trains, and equips space forces in order to protect US and allied interests in space and to provide space capabilities to the joint force. USSF responsibilities will include developing military space professionals, acquiring military space systems, maturing the military doctrine for space power, and organizing space forces to present to our Combatant Commands.

### Air Force Research Laboratory | afresearchlab.com/technology/space-vehicles/
The Air Force Research Laboratory's mission is leading the discovery, development, and integration of warfighting technologies for our air, space and cyberspace forces. With its headquarters at Kirtland Air Force Base, N.M., the Space Vehicles Directorate serves as the Air Force's "Center of Excellence" for space research and development. The Directorate develops and transitions space technologies for more effective, more affordable warfighter missions.

### Defense Innovation Unit | diu.mil
The Defense Innovation Unit's (DIU) mission is to accelerate commercial innovation for national security. It does so by increasing the adoption of commercial technology throughout the military and growing the national security innovation base. DIU's Space Portfolio facilitates the Department of Defense's ability to access and leverage the growing commercial investment in new space to address existing capability gaps, improve decision making, enable a shared common operating picture with allies, and help preserve the United States' superiority in space.

**DISTRIBUTION STATEMENT A**. Approved for public release: distribution unlimited.

**Cover:** Illustration depicting on-orbit servicing (Source: SkyCorp)

# TABLE OF CONTENTS

Foreword . . . . . . . . . . . . . . . . . . . . . . . . . . . . . . . . . . . . . . . . . . . . . . . . . . . . . . . . . . . . . . . . . iii
Executive Summary . . . . . . . . . . . . . . . . . . . . . . . . . . . . . . . . . . . . . . . . . . . . . . . . . . . . . . 1
Introduction . . . . . . . . . . . . . . . . . . . . . . . . . . . . . . . . . . . . . . . . . . . . . . . . . . . . . . . . . . . . 5
Space Policy & Financial Tools . . . . . . . . . . . . . . . . . . . . . . . . . . . . . . . . . . . . . . . . . . . 17
Space Information Services . . . . . . . . . . . . . . . . . . . . . . . . . . . . . . . . . . . . . . . . . . . . . 23
Space Transportation & Logistics . . . . . . . . . . . . . . . . . . . . . . . . . . . . . . . . . . . . . . . . 29
Human Presence . . . . . . . . . . . . . . . . . . . . . . . . . . . . . . . . . . . . . . . . . . . . . . . . . . . . . . 35
In-Space Power . . . . . . . . . . . . . . . . . . . . . . . . . . . . . . . . . . . . . . . . . . . . . . . . . . . . . . . 39
Space Manufacturing & Resource Extraction . . . . . . . . . . . . . . . . . . . . . . . . . . . . . 45
Insights & Recommendations for Industry . . . . . . . . . . . . . . . . . . . . . . . . . . . . . . . 51
Summary & Conclusions . . . . . . . . . . . . . . . . . . . . . . . . . . . . . . . . . . . . . . . . . . . . . . . 53
Appendix A . . . . . . . . . . . . . . . . . . . . . . . . . . . . . . . . . . . . . . . . . . . . . . . . . . . . . . . . . . A-1
Appendix B . . . . . . . . . . . . . . . . . . . . . . . . . . . . . . . . . . . . . . . . . . . . . . . . . . . . . . . . . . B-1
Appendix C . . . . . . . . . . . . . . . . . . . . . . . . . . . . . . . . . . . . . . . . . . . . . . . . . . . . . . . . . . C-1
Appendix D . . . . . . . . . . . . . . . . . . . . . . . . . . . . . . . . . . . . . . . . . . . . . . . . . . . . . . . . . . D-1
Appendix E . . . . . . . . . . . . . . . . . . . . . . . . . . . . . . . . . . . . . . . . . . . . . . . . . . . . . . . . . . E-1

# FOREWORD

A secure, stable, and accessible space domain is a vital interest to the United States. With the standup of the United States Space Force, we have a clear mandate from the highest levels of national leadership to ensure the many elements of space that have become integral to our way of life are protected now and into the future. Today, America is the best in the world at space and our adversaries know it. But leadership in space is not a right, and nations like Russia and China are racing to catch up to our capabilities and deny the benefits we derive from operations in, to, and from space.

The advantage America enjoys today stems from our space industrial base. We must work together to ensure that it remains a strong, effective, and innovative partner in sustaining American space superiority. The 2020 Defense Space Strategy recognizes that commercial space activities have expanded significantly in both volume and diversity. The Space Force must leverage the technologies and approaches at the frontier of this commercial expansion and move quickly to both reap the benefits of improving capability and affordability, and to be an enabling partner for American industry. By working with commercial partners, we will harness the best of both civil and government technology to further accelerate capabilities and expand the overall space economy.

The 2020 State of the Space Industrial Base Workshop brought together more than 120 voices from across the federal government, industry, and academia to assess the current health of the space industry and to provide recommendations for strengthening that industrial base. While the findings and recommendations from that workshop do not represent the official position of the United States Space Force, or any other branch of the government, it is important that we listen to these insights and evaluate the feasibility of implementing them in the advancement of national interests. America's future in space is a partnership and, as with any partnership, communication is key.

GENERAL JOHN W. RAYMOND
Chief of Space Operations, United States Space Force

*This page left intentionally blank.*

# EXECUTIVE SUMMARY

*"Support for a vibrant domestic manufacturing sector, a solid defense industrial base, and resilient supply chains is a national priority."*

National Security Strategy of the United States (2017)[1]

The world stands at the threshold of a new era in space. Much of the foundation of this exciting era is American made, and much of the innovation powering it is born of American creativity and ingenuity. Other nations, however, are challenging the US for leadership of this next space age. Success in this long-term strategic competition requires that the US seamlessly integrate multiple elements of national power. This report provides US policymakers and industry leaders comprehensive recommendations on the path

*Electron Rocket launches from the Mahia Peninsula (Source: Rocket Lab USA).*

forward to address the growing threats to US space power and how to ensure a strong US space industrial base as a foundation to US space leadership. Going beyond "admiring the problem," this report documents the next higher level of analysis of the challenge, focusing on specific actions the US must undertake to maintain security in space. The report clearly delineates the state of the space industrial base, its challenges and issues, and provides six overarching recommendations to US policymakers and four overarching recommendations to space industry leaders. It further provides detailed analysis of sub-areas of the space industrial base and areas affecting the base, specifying 39 actions as to what must be done, who must do them, and the timeframe in which they must be accomplished. Consistent with the National Defense Strategy[2] (NDS), it provides recommendations of how the Department of Defense (DoD) can integrate with the interagency to employ all dimensions of national power, and how it can create a 21st century national security innovation base that advances our security and economic strength. In June 2020, the new Defense Space Strategy (DSS) identified multiple lines of effort for US defense and military competitive advantage over foreign adversaries.[3] In July 2020, the National Space Council outlined its strategy for "low Earth orbit (LEO) commercialization, robotic and human exploration, national security capabilities, and international cooperation for science, safety, security, and economic growth" and the need for a

---

[1] United States. *The National Security Strategy of the United States of America*. Washington D.C: The White House. December 2017. https://www.whitehouse.gov/wp-content/uploads/2017/12/NSS-Final-12-18-2017-0905.pdf
[2] United States. *Summary of the 2018 National Defense Strategy of the United States of America*. Washington D.C: US Department of Defense. https://dod.defense.gov/Portals/1/Documents/pubs/2018-National-Defense-Strategy-Summary.pdf
[3] United States. *Defense Space Strategy Summary*. Washington D.C.: Department of Defense. 17 June 2020. https://media.defense.gov/2020/Jun/17/2002317391/-1/-1/1/2020_DEFENSE_SPACE_STRATEGY_SUMMARY.PDF

"whole-of-government approach to [US] space activities."[4] This report operationalizes both lines of effort by offering the next level of strategic analysis and required actions.

The findings and recommendations are the collective wisdom of over 120 space leaders from across government, industry and academia who assembled for the "State of the US Space Industrial Base 2020 Conference and Workshop," hosted by New Space New Mexico between 4-7 May 2020. This effort draws on and culminates three prior workshops to assist development of a comprehensive unified civil, commercial and national security space strategy for "maintaining and advancing United States dominance and strategic leadership in space."[5] This report does not constitute an advocacy position of the sponsoring agencies, or US government endorsement, recommendation, or favoring, but accurately documents the findings and recommendations of the conference.

The conference sponsors, the United States Space Force (USSF), the Defense Innovation Unit (DIU), and the United States Air Force Research Laboratory (AFRL), organized the workshop around the six areas vital to overall US national space power and the US space industrial base, and the areas most likely to be centers of gravity in great power competition:

- **Space policy and finance tools** to secure US space leadership now and into the future by building a unity-of-effort and incentivizing the space industrial base.
- **Space information services** include space communications/internet, positioning, navigation and timing (PNT), and the full range of Earth observing functions which have commercial, civil and military applications.
- **Space transportation and logistics** to, in and from cislunar space and beyond.
- **Human presence** in space for exploration, space tourism, space manufacturing and resource extraction.
- **Power for space systems** to enable the full range of emerging space applications.
- **Space manufacturing and resource extraction** for terrestrial and in space markets.

For each area, the participants analyzed its current state, challenges, and inflection points and, based on that analysis, developed focused recommendations, consistent with national space policy as to:

- US government policy and actions to maintain and expand the US space industrial base required for national security needs.
- Whole-of-government policies and actions to guarantee the US space industrial base for a future with space as a major contributor to US national power.
- Immediate actions to preserve and expand the US space industrial base in light of the disruption by the Coronavirus Disease 2019 (COVID-19) pandemic.

From these efforts, the working groups and sponsoring organizations identified the specific actions necessary to address the challenges identified in each area. These are summarized in the report. In addition, they identified the most important actions for government and industry that cut across these

---

[4] United States. *A New Era for Deep Space Exploration and Development*. Washington D.C.: National Space Council. 23 July 2020. https://www.whitehouse.gov/wp-content/uploads/2020/07/A-New-Era-for-Space-Exploration-and-Development-07-23-2020.pdf.

[5] United States. Compilation of Presidential Documents. *DCPD-201900558-National Security Presidential Memorandum on the Launch of Spacecraft Containing Space Nuclear Systems*. Washington D.C: Office of the Federal Register, National Archives and Records Administration. 20 August 2019. https://www.govinfo.gov/app/details/DCPD-201900558

areas. These actions lie at the nexus of shared interests of the DoD, whole-of-government, and the broader industrial base. These are discussed in the detail in the report but can be summarized for government as:

1. The US government promulgates a whole-of-government, "North Star" top-level vision and strategy for space industrial development and establishes a Presidential Task Force to execute it.
2. DoD develops plans to protect, support, and leverage commerce in space.
3. The US government works to economically stimulate the industry, including space bonds and a Space Commodities Exchange and by executing $1 billion of existing DoD and NASA funding through the Exchange.
4. The US government develops a framework for creating wealth and security with allies and partners that share our common norms and values.
5. The US government supplies the workforce necessary to fill more than 10,000 Science Technology Engineering and Math (STEM) jobs domestically.
6. The USSF works closely with space industry entrepreneurs and innovators to develop government-commercial technology partnerships that support US commerce and national security in space.

Key recommendations for industry detailed in the last section of the report are summarized below:

1. Industry should aggressively pursue partnerships with the US government to develop and operate joint commercial, civil and defense space capabilities. These partnerships should jointly fund developing capabilities that benefit from but are not heavily reliant on US government investment and revenue for their commercial viability.
2. Entrepreneurs with innovative and potentially dual-use technologies must improve the protection of their intellectual property from unintended foreign assimilation, including protecting their networks from cyber exfiltration attempts, and avoiding exit strategies that transfer intellectual property to foreign control hostile to US interests.
3. Businesses should engage across the US educational system to guide and develop the future STEM workforce to fuel the future space economy, to include funding undergraduate scholarships/loans for STEM

*US essential aerospace technicians wearing personal protective equipment continue assembly of the Dream Chaser Cargo Module during the COVID-19 pandemic (Source: Sierra Nevada Corp).*

students, internships and providing space professionals to support instruction in space subjects.

4. Industry should improve ties and partnerships with domestic and allied parts, subcomponent and subsystem manufacturers to strengthen trust and resilience in space supply chains.

# INTRODUCTION

In 2019, the US Air Force Space Command assessed that by 2060 space will be "a significant engine of national political, economic, and military power" and that "the United States must commit to having a military force structure that can defend this international space order and defend US space interests, to include US space settlements and commerce."[6] The United States can either prepare and posture to shape a future with US strategic leadership in space, or resign itself to second class status.[7]

Private investment in commercial space activity is at an all-time high and growing as entrepreneurs and industrialists create new technologies and adapt existing technologies for space application. This is fueled by the decreasing cost of space access and broad advances in space enabling technologies. This provides the opportunity for an expanded space industrial base beyond "big" aerospace companies that have traditionally supported government space missions. Much like the early days of the automobile or aviation, increased innovation is being driven by the addition of these independent visionaries who seek to find new, efficient, and effective ways to make money in space. These "new space" entrants are a fast-growing segment of the US space industrial base and a key element of the report is aimed at recommendations and strategies that can enhance this element of the US space industrial base.

Over the last twenty years, there has been ongoing recognition by the government of the growing importance of space to national power and the increasing challenges to US preeminence. At the turn of the century the Congressional Commission of 2000[8] and the Presidential Commission of 2002[9] took a holistic look at the issues and actions required. Unfortunately, limited progress was made in implementing their findings due to shifts in national focus and priorities to counterterrorism in the wake of the September 11 attacks. More recently there has been significant progress with the revival of the National Space Council, and the promulgation of a new National Space Strategy in 2018, that "emphasizes dynamic and cooperative interplay between the national security, commercial, and civil space sectors," and the release of the National Space Council's vision for space exploration and development in July 2020.[10] This forward momentum, among other policy successes, resulted in establishing the new USSF, reinvigorating NASA's space programs, transferring space traffic management responsibility, and modernizing commercial space regulations. In June 2020, the new DSS identified multiple lines of effort for US defense and military competitive advantage over foreign

---

[6] Mozer, Joel, Dr. *The Future of Space 2060 and Implications for U.S. Strategy: Report on the Space Futures Workshop*. Air Force Space Command, September 5 2019. 1-32. (See Appendix B).

[7] Sadat, Mir, Dr. *"Space Cooperation in an Age of Great Power Competition in the Indo-Pacific."* Hudson Institute. 2019. https://www.youtube.com/watch?v=63B74CEbpsg

[8] U.S. Congress. *The Commission to Assess United States National Security Space Management and Organization.* Public Law 106-65, the National Defense Authorization Act for Fiscal Year 2000, Section 1622. https://www.congress.gov

[9] United States. *Final Report of The Commission On The Future Of The United States Aerospace Industry.* Washington: Commission on the Future of the United States Aerospace Industry. November 2002. https://history.nasa.gov/AeroCommissionFinalReport.pdf

[10] Smith, Marcia. *"White House Releases Fact Sheet On New National Space Strategy."* SpacePolicyOnline.com. 24 March 2018. https://spacepolicyonline.com/news/white-house-releases-fact-sheet-on-new-national-space-strategy/; United States. *A New Era for Deep Space Exploration and Development.* Washington D.C.: NAtional Space Council. 23 July 2020. https://www.whitehouse.gov/wp-content/uploads/2020/07/A-New-Era-for-Space-Exploration-and-Development-07-23-2020.pdf.

adversaries. This report operationalizes the DSS lines of effort by offering the next level of strategic analysis and required actions.

Beyond this progress, much work remains to create and execute the integrated, comprehensive national space strategy that synchronizes national security, civil and commercial space efforts to ensure a 2060 American Space Vision; a vision to enhance American's national defense space capabilities, to create profitable new space industries, to develop space communities, and to use space to protect and enhance life on Earth; all these as part of a unique cislunar economy enabling human expansion into the Solar System.

## MOTIVATION

Space is again a focal point of renewed great power competition.[11] The once significant lead enjoyed by the US in space is challenged as the barriers of space access have lowered and US technological superiority has eroded. A global commercial industrial and talent base now is broadly proliferating space access, engineering design skills and manufacturing capabilities. Without an integrated, comprehensive national space vision and strategy, US space leadership and competitive advantage are at risk. The *2017 US National Security Strategy* calls for advancing space as a priority domain, promoting space commerce, and maintaining the US lead in space exploration.[12] A recurring theme in US policy is "maintaining and advancing United States dominance and strategic leadership in space."[13] To support continued space dominance, the USSF, DIU, and AFRL sponsored government-industry engagements to examine the state of the US space industrial base and identify the crucial national policies and actions to fully employ all relevant instruments of national power to ensure US space power.[14] The three sponsors of this report, the USSF, DIU, and AFRL, cognizant of the challenges before our nation, sought to capture the current state of the US space industrial base and identify the specific

> *"It's our destiny to explore. It's our destiny to be a space-faring nation. We leave as we came and, God willing, as we shall return, with peace, and hope for all mankind."*
>
> Eugene Cernan, Apollo 17 Astronaut

*Captain 'Gene' Cernan (USN) was the last man to walk on the Moon in 1972 (Source: NASA)*

---

[11] Patrick M. Cronin, P., Murano, M. & H.R. McMaster. "Transcript: US Space Strategy and Indo-Pacific Cooperation." The Hudson Institute. 15 November 2019.
https://www.hudson.org/research/15481-transcript-u-s-space-strategy-and-indo-pacific-cooperation
[12] United States. *The National Security Strategy of the United States of America*. Washington D.C: The White House. December 2017. https://www.whitehouse.gov/wp-content/uploads/2017/12/NSS-Final-12-18-2017-0905.pdf
[13] United States. Compilation of Presidential Documents. *DCPD-201900558-National Security Presidential Memorandum on the Launch of Spacecraft Containing Space Nuclear Systems*. Washington D.C: Office of the Federal Register, National Archives and Records Administration. 20 August 2019. https://www.govinfo.gov/app/details/DCPD-201900558
[14] Traditional and emerging instruments of national power include diplomatic, information, military, economic, financial, legal (law enforcement), scientific and technological, and environmental.

short-, mid- and long-term actions to ensure the US space industrial base required to maintain US leadership in space as an expanding domain of human and commercial activity, and as a source of US national power. The conference was also undertaken to address the clear and present danger the COVID-19 pandemic presented to the space industrial base as detailed in Appendix D.

## ASSEMBLING AN INTERDISCIPLINARY TEAM OF EXPERTS

The findings and recommendations are the collective wisdom of over 120 space leaders from across government, industry and academia who assembled for the "State of the US Space Industrial Base 2020 Conference and Workshop," hosted by New Space New Mexico from 4-7 May 2020. This effort drew upon and culminated three prior workshops held to support development of a comprehensive unified civil, commercial and national security space strategy. Previous efforts[15] defined the range of possible futures, the desired futures advantageous to the US, and the concerns as to the state of our industrial base. These provided the foundations defining where we are and where we wish to go. This effort built upon those efforts to define the actions by which the US can accomplish its objectives.

## BEYOND ADMIRING THE PROBLEM

This report provides policymakers comprehensive recommendations to address the growing threats to US space power and to ensure the strong US space industrial base foundational to US space leadership. Going beyond "admiring the problem," it documents the next level of analysis of the challenge, focusing on specific actions the US must undertake to maintain security in space. The report clearly delineates the state of the space industrial base, its challenges and issues, and provides six overarching recommendations to US policymakers and four overarching recommendations to industry partners. It further provides detailed analysis of six sub-areas of the space industrial base, specifying 39 actions as to what must be done, who must do them, and the timeframe in which they must be done.

## THE CENTERS OF GRAVITY FOR SPACE LEADERSHIP

The sponsors, USSF, DIU, and AFRL, in support of this aim organized the workshop around the six areas most vital to over US national power in space, and the areas most likely to be at the center of gravity in great power competition:

- **Space policy and finance tools** to secure US space leadership now and into the future by building a unity-of-effort within the government and incentivizing the space industrial base.
- **Space information services** including space communications/internet, PNT, and the full range of Earth observing functions which have commercial, civil and military applications.
- **Space transportation and logistics** to, in and from cislunar space and beyond.
- **Human presence** in space for exploration, space tourism, space manufacturing and resource extraction.
- **Power for space systems** to enable the full range of emerging space applications.
- **Space manufacturing and resource extraction** for terrestrial and in space markets.

---

[15] See Appendix B for links to prior workshop reports.

## FOCUSED ANALYSIS

For each area, the participants analyzed its current state, challenges, and inflection points and based on that analysis developed focused recommendations, consistent with national space policy,[16] as to:

- US government policy and actions to maintain and expand the US space industrial base required for national security needs.
- Whole-of-government policies and actions to guarantee the US space industrial base for a future with space as a major contributor to US national power.
- Immediate actions to preserve and expand the US space industrial base in light of the disruption by the COVID-19 pandemic.[17]

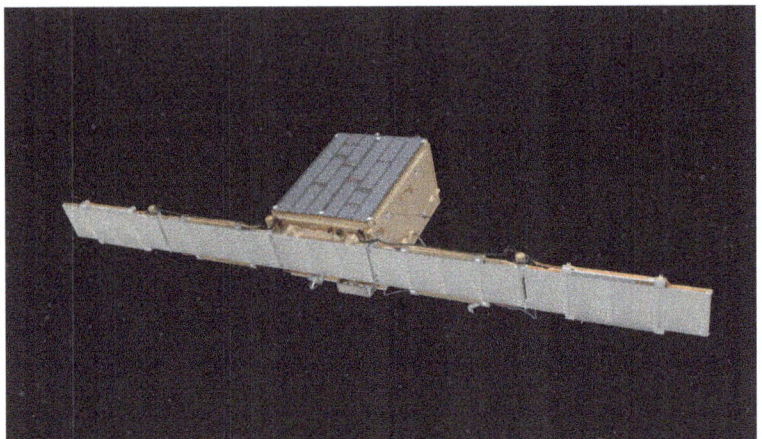

*An X-Band Synthetic Aperture Radar (SAR) smallsat utilizing an active phased array antenna (Source: R2 Space)*

Working group leadership and sponsors identified those actions most essential to catalyze broad adaptive change and contribute most effectively to the maintenance and expansion of US leadership in space. These actions lie at the nexus of shared interests of the DoD, and the broader industrial base, where the sponsoring agencies play a role.

## REPORT OVERVIEW

The report begins with a summary of general observations of the state of US space power and ambitions and the current challenges the US faces in maintaining and expanding US commercial, civil and military space leadership. The report next details general recommendations for actions paramount to advancing space as a priority domain, promoting space commerce, maintaining leadership in space exploration, and sustaining overall US security and military advantage in the space domain. The remaining sections summarize the challenges, inflection points and 39 recommendations for specific actions across the six sectors to ensure continued US leadership and security. A last section covers observations and recommendations for industry action to strengthen the US space industrial base.

## GENERAL OBSERVATIONS AND RECOMMENDATIONS

Workshop participants represented leaders from across the space community. The participants worked within the six topical areas and each group produced an initial report on their findings and recommendations summarized later in this report. In discussion between the groups and upon reviewing their findings and recommendations, overarching observations, and recommendations were

---

[16] United States. *The National Security Strategy of the United States of America*. The White House. 2017.
[17] See Appendix D for COVID-19 specific impacts to the US space industrial base.

determined that span the six areas. These build upon the central observation that through a powerful set of interdependent and mutually reinforcing civil, commercial and national defense capabilities, the US can maintain and responsibly unleash its space power to ensure its dominant position consistent with our core democratic norms and values. Key to achieving this is a robust space industrial base.

The remainder of this section summarizes the overarching status, challenges and recommendations toward achieving that powerful set of interdependent and mutually reinforcing civil, commercial, and national security capabilities across the explored six areas.

## CURRENT STATE

The United States is well positioned to achieve a future of increased space power consistent with the long-standing doctrine of "Peace through Strength."[18]

**The US remains a space civil, commercial, and national security leader** - No nation has yet matched the US accomplishments in space exploration. The US is the first nation to demonstrate commercial orbital cargo delivery, commercial heavy lift, commercial first-stage reusability, deployment of space-based mega-constellations for overhead sensing, internet broadband, and commercial human spaceflight. The US military maintains the most capable military constellation including unique capabilities such as the X-37B spaceplane.

*A Falcon 9 reusable first stage during hypersonic reentry and landing at Cape Canaveral, Florida on 25 July 2019. The same core was flown on 4 May 2019 - 82 days prior - and then refurbished. (Source: SpaceX)*

**There is a renewed national focus on space across civil, commercial, and national security domains** - Increased awareness of the criticality of US space capabilities is driving a national sense of urgency for whole of government alignment and resourcing. The US is pursuing aggressively a return to the Moon by 2024 and the development of the capabilities and infrastructure for a sustained lunar presence as an intermediate step to Mars and beyond. There is growing recognition on the part of the Administration, Congress, NASA and DoD that a key element for continued, long-term, US economic and strategic dominance is to catalyze and enable the accelerated growth of a vibrant US private industrial and cultural expansion into the Solar System. Increased human presence and action in space

---

[18] Phrase attributed to the Roman Emperor Hadrian (AD 76–138). George Washington referred to such a policy in his 1793 State of the Union Address.

and its evolution from temporary to permanent requires continued US technological and exploration leadership. Bipartisan congressional efforts have authorized a Space Force to secure our national interests in this increasingly contested domain.[19]

**There is a concomitant shift in societal ambition** - New private investment in space led by space entrepreneurs such as Jeff Bezos, Elon Musk, Sir Richard Branson, and others are leading to breakthroughs in cost, efficiency, and creative solutions across the space field. The entrepreneurs are driven by their shared belief in the expansion of humanity into space and the eventual settlement of the solar system. Bezos dreams of "moving all heavy industry off Earth and Earth will be zoned residential and light industry" and a world for his great-grandchildren's grandchildren where humanity moves out into the solar system."[20] Musk aspires to colonize Mars and make humanity a multi-planetary species.[21] Some may question their idealism in favor of other rationales, but their impact on US' and allies' space technology advances and US commercial space advantage continues to be dramatic - even more so their impact on general public perception and enthusiasm for the future of space and for continued US space leadership.

*On 9 May 2019, Blue Origin founder, Jeff Bezos, announced Blue Moon - large lunar lander capable of delivering multiple metric tons of payload to the lunar surface based on configuration and mission. (Source: Blue Origin)*

**Increased partnering** - At every level of policy guidance, the DoD is pushing to increase partnerships to leverage commercial space technology to meet the need to move quickly and maintain affordability. Early successes include the Space Enterprise Consortium (SpEC) already used to solve some of the most

---

[19] Edmondson, Catie. "House Passes $738 Billion Military Bill With Space Force and Parental Leave." The New York Times. 11 December 2019. https://www.nytimes.com/2019/12/11/us/politics/house-ndaa-space-force-leave.html
[20] Clifford, Catherine. "Jeff Bezos Dreams of a World with a Trillion People Living in Space" CNBC. 1 May 2018. https://www.cnbc.com/2018/05/01/jeff-bezos-dreams-of-a-world-with-a-trillion-people-living-in-space.html
[21] Statt, Nick. "Elon Musk Says the Only Reason He Wants to Make Money Is to Colonize Mars" The Verge. 27 September 2016. https://www.theverge.com/2016/9/27/13079472/elon-musk-mars-space-x-tesla-funding-dream

complex defense space problems, the Space Pitch Day that award of $1.5 million to accelerate new technology, Catalyst Campus, SURF Camp, and SDA's SPACE Camp, and the National Security Space Launch (NSSL) which is helping create new engines and launch vehicles. These are important steps. Continuing calls from senior US government leaders to "go faster," can be met in part by expanded use of new commercial space transportation and space logistics capabilities, and extending Acquisition Demonstration Project authority to the USSF.

## ISSUES & CHALLENGES

Though well positioned, the US faces significant challenges. These are:

**Contested leadership** - The US is not alone in planning to return humans to the Moon or expanding the use of space. China has announced its intention to do so by 2035.[22] China is committed and credible in its pledge to become the leading, global super-power, to include space, by 2049 marking the 100th anniversary of the People's Republic. A key component of China's strategy is to displace the US as the leading power in space and lure US allies and partners away from US-led space initiatives, through its Belt and Road initiative and plans for an Earth Moon Economic Zone.[23] Nor is China alone. An expanding number of nations are emerging as space competitors and potential adversaries. Meanwhile, the US lacks a national-level "North Star" vision for space industrial and economic development to align efforts across the space sector to maintain US leadership.

**Expansion of activity across cislunar space and control of critical choke points** - As space activities expand beyond geosynchronous orbit, the first nation to establish transportation infrastructure and logistics capabilities serving GEO and cislunar space will have superior ability to exercise control of cislunar space and in particular the Lagrange points and the resources of the Moon. Lunar resources, including hydrogen/oxygen for propellant that enable cheaper mobility for civil, commercial and national security applications, are key for access to asteroid resources and Mars, and to enable overall space commercial development.

**Expanding the role of the USSF** - An increasing US presence within an expanded cislunar economy will require security and a stabilizing military presence. Clarifying the USSF role in protecting and enabling US commerce across cislunar space is critical. As we emerge from the period of strategic redefinition and changing focus, the USSF can address our eroding competitive military advantage and reinforce our traditional tools of diplomacy. This will enhance the US ability to negotiate from a position of strength, ensure the balance of power remains in our favor, and advance an international order that supports our security and prosperity. A Space Force that presents forces, capabilities, and doctrine for national space power with the clearly defined mission to enable and protect US space commerce will project confidence and lower the perception of risk by providing stabilizing presence, surveillance, aids to navigation, and help when required.

---

[22] Wall, Mike. "China Just Landed on the Moon's Far Side - and Will Probably Send Astronauts on Lunar Trips." space.com. 5 January 2019. https://www.space.com/42914-china-far-side-moon-landing-crewed-lunar-plans.html
[23] Siqi, Cao. "China Mulls $10 Trillion Earth-Moon Economic Zone." Global Times. 1 November 2019. https://www.globaltimes.cn/content/1168698.shtml

*Three Planet SkySat imaging satellites deploy ahead of 58 StarLink broadband satellites on June 13th, 2020. (Credit: SpaceX)*

**Damage from the impact of COVID-19** - The COVID-19 pandemic has impacted the entire US space industry. The magnitude and success of the industry's recovery is uncertain and unpredictable (See Appendix D). Maintenance and growth of the US space industrial base requires the return of early-stage investment in launch, services, manufacturing and logistics companies. For the US, the COVID-19-produced global recession provides a dangerous opportunity for near-peer competitors to challenge or surpass the US in Space by 1) strategic acquisitions of companies and their intellectual property (IP) and/or 2) maintaining investments while the US private capital is unavailable and the US government is distracted. Under the economic stress of COVID-19, the present levels of venture capital driving US space innovation will likely taper off in the absence of certainty, leaving the primary driver of space development to again be the US government. In addition, over the last 5 years, $11 billion of private non-venture capital has been invested.[24] The current, limited return on investment (ROI) on this early investment in space may drive out further investment. To cope, the US will need to employ a full set of offensive and defensive financial economic tools to attract investor diversity, increase successful exits and normalize space as an extension of the global (terrestrial) economy.

**Flat defense budgets** - DoD faces growing capabilities and threats from our adversaries. Future DoD budgets will most likely be flat or declining, in part exacerbated by COVID-19. This will strain DoD's ability to develop significant, new DoD Space systems alone. This will place a premium on DoD's leveraging of commercially developed systems to the maximum extent possible, working within the capabilities of those systems, rather than procuring more capable, but less affordable capabilities.

**Complex government procurement processes** - US procurement practices are complex and often opaque to the private sector. This limits the ability of the US government to send a clear and predictable demand signal to the private sector and leverage its market power to drive down costs. DoD

---

[24] See Appendix D.

space programs, budgets, requirements and acquisition processes remain largely unchanged. Government procured space systems historically are characterized by high cost, long program schedules and frequent delays. This has allowed potential foreign adversaries to develop their space programs at a faster pace than the US, putting the US at a strategic disadvantage. Technological and procedural advances occurring in other economic sectors have not made it into space programs. Failure to consolidate DoD satellite communication services procurement within a single service or force hampers efficient DoD exploitation of present and emerging commercial space communication systems.

**Government legal and regulatory requirements** - US space companies face complex legal and regulatory requirements. For non-defense industrial base companies, the compliance requirements are often overwhelming and discourage new entrants from developing solutions for national security space purposes.

**Low rate of DoD space experimentation and prototype development** - The rapid advances of other nations' space capabilities challenge America's technological edge. Critical to retaining US technical leadership is a vigorous program of in-space experimentation and prototype development. DoD efforts in this area are underfunded and insufficiently coordinated across the department. The very long schedules of many DoD space programs limit the opportunities for short design-build-test projects that stimulate interest and innovation within the US workforce. Many new small launch vehicles are coming into service. An increased rate of space experiments and prototypes by DoD would enhance the viability of these small launchers providers. Choosing the role of a lighthouse customer could strongly influence the quantity and quality of commercial offerings. Extending the USAF's Acquisition Demonstration Project Authority to the USSF is critical to maintaining a fast pace of innovation[25] to introduce unpredictability to adversary decision-makers called for in the NDS.

*Air Force Space Pitch Day is part of an effort to demonstrate a faster, smarter strategy in technology investments and partnerships with small businesses. (Source: USAF)*

**STEM shortages** - The current and projected level of STEM personnel is insufficient to support the development of expanded national space capabilities within the US space industrial base. NASA's Artemis program will require an additional 10,000 STEM graduates for civil needs alone, with more needed to support the new USSF and an enlarged private commercial sector. In addition to STEM, the industry will require an increased number of non-STEM personnel knowledgeable of the space enterprise in a variety of support occupational fields such as financial engineering, economics and law. Avoiding such a deficit will require a whole-of-government mobilization.

**Architectural clarity** - Industry requires greater clarity from the DoD as to their long-term architectural needs and requirements and the roles and responsibilities of the USSF to protect lines of

---

[25] An authority contained in 2020 USSF legislative proposal Sec. 934.

commerce. The new USSF can stimulate the industrial base and enable a future pipeline of innovation as a 'lighthouse customer' acquiring capabilities within an architecture which incorporates new industrial markets and capabilities within an expanded space commercial economy. Such work is manpower and brainpower intensive, requiring the Chief of Space Operations to have a staff with the authority and composition commensurate to the task.[26]

## KEY ACTIONS & RECOMMENDATIONS

From the above observations and additional inputs from the working groups, participants determined six overarching recommendations for action:

**Create a national "North Star" space vision** - Building on the foundation for continued US space leadership created by recent policy and organizational advances, the US should develop a guiding national vision for long-term, space industrialization and national space development to catalyze whole-of-nation efforts and enable the United States to compete and win now and into the future. This should be developed to drive a host of federal department and agency specific actions and international partnerships. The steps to construct and rapidly implement such a vision are detailed in the Space Policy and Finance Tools section and Power for Space Systems section.

**Define the USSF roles and missions to protect and leverage US commercial and civil space capabilities** - The creation of a USSF is exciting and historic. As an innovative, fast and future looking force, it is tasked to provide forces to protect the interests of the United States in space; deter aggression in, from, and to space; and conduct space operations. As part of its mission, the USSF should articulate its role to secure commerce and protect civil infrastructure in the space domain. This examination should consider the degree to which this role should emulate the US Navy role in assuring the maritime domain. Clarity on this issue will drive commercial confidence for a more rapid expansion of US space entrepreneurial activity. When implementing this part of its mission, the USSF should examine an increased role in America's return to the Moon (such as providing safety of navigation services) and expanded opportunities for partnerships with companies to develop prototypes, to procure operational product services, and to sponsor new competition. The USSF should articulate its role in planetary defense. Such a role could accelerate America's edge in asteroid mining and in-space transportation. With a role similar to an Army Corps of Engineers, SeaBees or Coast Guard, the USSF can guide and accelerate the development of critical infrastructure.

**Develop and deploy new government tools to support growth of the US space industrial base** - Since space is a strategic industry critical to great power competition, the space industry must be encouraged and energized in the specific ways that ensure its needed contribution to national power. A broader set of tools available to the American people should be developed and deployed to nurture new capabilities and markets to include bonds, a space commodities exchange, and government commitment to procure products and services through such an exchange. Additional discussion of these tools are found in the Space Policy and Finance Tools section.

**Develop a wealth creation framework for like-minded allies & partners** - Present day great power competition in space is not principally about prestige or ideology signaling but about building powerful and enduring alliances and partnerships. A central challenge for the US is defeating our

---

[26] An authority contained in 2020 USSF legislative proposal Sec. 931

adversaries' efforts to pry away our allies and partners through offers of joint participation in the development of global platforms and international infrastructure and wealth, including space development. To reverse the weakening of the post-WWII order, the US should deepen our ties with allies and partners who share our vision of a free and open space domain through creating a meaningful alternative that moves beyond space exploration or military cooperation and provides a path toward prosperity from an expanded space economy secured through the stabilizing presence of the USSF.

**Mobilize STEM** - Successful competition is simply not possible without the human capital underlying our space innovation base. The US government must signal the importance of STEM, and must emplace incentives to fill the talent pipeline. This should include as a specific potential solution a "STEM ROTC," with targeted undergraduate scholarships for US citizens in return for working in STEM in the US after graduation.

These six recommendations are summarized below along with the organizations which have authority to implement them:

1. US promulgate a "North Star" top-level vision and strategy for space industrial development and establish a Presidential Task Force to execute it (OPRs: POTUS, VP, NSpC, NSC, NEC, OTMP);
2. US Department of Defense develops plans to protect, and support and leverage commerce in space (OPRs: OSD, USSF, USSPACECOM);
3. US government work to economically stimulate the industry, including space bonds and a Space Commodities Exchange and by executing $1 billion of existing DoD and NASA funding through the Exchange (OPRs: NSpC, NEC, OTMP, DOC, SBA);
4. US government develop a framework for creating wealth and security with allies and partners that share our common norms and values (OPRs: NSpC, NSC, NEC, DOS, DOC);
5. US government supply the workforce necessary to fill more than 10,000 Science Technology Engineering and Math (STEM) jobs domestically (OPRs: NSpC, OSTP, DPC, DOEd); and
6. USSF works closely with space industry entrepreneurs and innovators to develop government-commercial technology partnerships that support US commerce and national security in space (OPRs: USSF, DIU).

These recommendations will place the United States on a firm footing for the challenges it faces. The remaining sections provide analysis and specific recommendations across six areas. First, we examine specific policy and financial tools available to the United States to ignite the broader industry. Subsequently, we review the current state and necessary actions across the five critical subcomponents of the space industry which will determine the future of US security and prosperity.

An integrated, comprehensive national strategy must identify the priorities, aspirations, challenges, and actions required to maintain and expand the US space industrial base foundational to US civil, commercial, and military space capabilities. These recommendations are topline areas of emphasis which demand the immediate and enduring attention of legislative and executive branch policymakers. Our full report provides additional specifics.

*This page left intentionally blank.*

# SPACE POLICY & FINANCE TOOLS

*"The foundation of war is economics. If you have half the resources of the counterparty then you better be real innovative. If you're not innovative, you're going to lose."*

Elon Musk, Entrepreneur[27]

## BACKGROUND

Policy is central in the maintenance of the US national space industrial and innovation base. The renewed attention to space policy and strategy presents important opportunities to sustain and expand US space power. In support of our nation's ambitions, visionary space policies enable whole-of-nation alignment and attract new partners. By advancing interagency integrating policies, the US can synthesize a broad and diverse technological base, ensure US global competitiveness and leverage and protect ally and partner capabilities in a global marketplace advantageous to US interests. Sound fiscal and acquisition policies can minimize market risk, support the infrastructure foundational to technological innovation, and ensure US competitiveness in a global market. To realize these opportunities requires an enduring economic and national security policy regime that promotes and incentivizes growth and innovation of the space industrial base in concert with trusted allies and partners. Such a policy regime empowers the various US government agencies to deploy the full range of government financial resources and technical expertise to unleash the vibrant US commercial sector, and shape prudent norms, rules, and behavior for a favorable order in space.

## CURRENT STATE

US space policies historically focused on civil exploration and protection of national security information services shaped by its dominant position, the relatively small size of commercial space, and the limited number of competing spacefaring nations. The needs and scope of space policy is changing, driven by the continued expansion in space activities and the growth in numbers and the improved capabilities of ally and adversary spacefaring nations. In this renewed era of great power competition, space policy is moving to address the rapid growth in the US and global commercial space sectors, predatory practices by our competitors and the use of commercial space companies as state capital proxies by foreign adversaries. Space policy is moving to harness the full range of US economic and financial tools available or required to be developed to enable the US space industrial base. This includes policies and financial actions that address the availability of patient capital to finance the relevant hardware-intensive technologies and the presence of a robust market demand signal, either in the private sector or from the government. These are the central requirements for a burgeoning new space industrial base in the United States.

The renewed attention to policy and strategy is seen in the revival of the National Space Council (NSpC) in June 2017. The NSpC marks a tremendous step toward guiding the nation's topline space

---

[27] "Air Force Association. Elon Musk and Lt Gen JT Thompson. "YouTube Video." February 28, 2020. https://www.youtube.com/watch?v=E307nHamdY8

policies for our economy, national security, and scientific exploration. In February 2020, the membership of the National Space Council was amended to include the Secretary of Energy, the Assistant to the President for Economic Policy, and the Assistant to the President for Domestic Policy.[28] The NSpC, along with the National Security Council (NSC), have driven the birth of the new USSF, reinvigorated NASA's human space program, advanced space traffic management, streamlined commercial space regulations, and created the Space Acquisition Council to improve coordination and synchronization of DoD space systems and program acquisitions, among many other policy initiative successes.[29]

## KEY ISSUES & CHALLENGES

### POLICY

**Keeping pace with the rapid pace of change in space** - US space policy development and revision is challenged to keep up with changes in great power competition, dynamic changes globally in the commercial market and technological innovation.

**Export controls stifle US economic competitiveness** - The International Traffic in Arms Regulations (ITAR) on military-related technologies have unintentionally limited international collaboration, innovation and trade, preventing US space companies from being the suppliers of choice in a globally competitive economy. US companies' potential market shares have been eroded by the rise of non-ITAR companies in Europe.[30] The resulting environment has stifled even strategic efforts to collaborate with our closest allies.

**Lack of global supply chain transparency** - Foreign manipulation and irresponsible use of supply chains is a concern at the highest levels of the US government.[31] Complex, global, supply chains introduce a high degree of risk to domestic companies, and in turn, national security space end-users. The United States, along with its allies and partners, must understand the origin and assembly process of critical components for complex space-based technologies. Supply chain hygiene is crucial to the development of space technologies for US national security purposes and for commercial purposes.

---

[28] United States. *DCPD-202000074-Amending Executive Order 13803 – Reviving the National Space Council.* Vol 82, No. 239. Washington D.C: Federal Register Online via Government Publishing Office. 13 February 2020.

[29] United States. *82 FR 49501-Memorandum on National Space Traffic Management Policy.* Vol 82, No. 239. Washington D.C: Federal Register Online via Government Publishing Office Dec. 2017. https://www.govinfo.gov/content/pkg/FR-2017-12-14/html/2017-27160.htm; United States. *83 FR 24901- Memorandum Streamlining Regulations on Commercial Use of Space.* May 30, 2018. https://www.govinfo.gov/content/pkg/FR-2018-05-30/html/2018-11769.htm ; United States. *DCPD-201800431. Memorandum on National Space Traffic Management Policy.* 18 June 2018. https://www.govinfo.gov/content/pkg/DCPD-201800431/html/DCPD-201800431.htm; United States. *84 FR 6049-Memorandum on Establishment of the United States Space Force.* Feb. 25, 2019. https://www.govinfo.gov/content/pkg/FR-2019-02-25/html/2019-03345.htm Erwin, Sandra. "Space Acquisition Council to Consider New Actions to Help the Aerospace Industry Cope with Pandemic."SpaceNews. 11 April 2020. https://spacenews.com/space-acquisition-council-to-discuss-future-actions-to-help-suppliers-cope-with-pandemic/.

[30] Smith, Marcia. "Space Council Adopts Recommendations at its March 26, 2019 Meeting." SpacePolicyOnline. March 26, 2019.

[31] United States. *84 FR 22689-Executive Order on Securing the Information and Communications Technology and Services Supply Chain.* Washington D.C: Federal Register Online via Government Publishing Office. May 17 2019. https://www.govinfo.gov/content/pkg/FR-2019-05-17/html/2019-10538.htm

## FINANCE

**Developing and employing economic tools** - The US must develop policy that allows deployment and full utilization of existing economic offensive and defensive tools across the full spectrum of the space market. The US must also promote the development of new tools as needed to increase US commercial space activities and support the growth of US space companies.

**Loss/decreases of venture funding of the space innovation base** - Additional capital investments may dry up. In the last 5 years, $11 billion of private capital has been invested in commercial space technology companies, whose most promising end-users were in the US government.[32] The fiscal constraints exacerbated by COVID-19 spending, will dry up many sources of future government demand. The current lack of return on early investment in space, coupled with lack of consumption, could deter further investment. Without action, the expected post-COVID-19 global recession will dry up further investments into the space sector.

**Maintenance of Space R&D** - In the aftermath of the virus pandemic, traditional private investors may become reluctant to fund space technologies due to higher-risk and longer time horizons. Financing gaps between early-stage companies and commercialization, and an overall lack of liquidity for the space technology ecosystem also complicate financing within the new space industrial base.

**Government procurement complexities** - US procurement practices are complex, siloed, and largely opaque to the private sector. US space companies face complex legal and regulatory requirements. For non-defense industrial base companies, the compliance requirements are often overwhelming and discourage new entrants from developing solutions for national security purposes[33]. This complexity limits the ability of the US government to leverage its market power to drive down costs and send a clear and predictable demand signal to the private sector.

## KEY INFLECTION POINTS

### ENABLING INFLECTION POINTS

- **US achieves political and financial economic alignment** across government, industry, academia, and allies and partners to pursue an enduring strategy in commercial space.
- **US plays a central role in establishing international norms, behavior, rules and laws** for civil and commercial space.
- **Commodification of the space commerce industry** leads to cheaper goods and services due to innovation, increased demand, and new entrants into the market creating jobs and economic growth.

### LIMITING INFLECTION POINTS

- **Economic recession(s) following the COVID-19 pandemic** limits the ability of the US and other governments to support the space industry.

---

[32] See Appendix D.
[33] Williams, Lauren. "Why Congress Holds the Key to DoD Tech." FCW. 6 February 2020. https://fcw.com/articles/2020/02/06/dod-tech-randd-congress.aspx

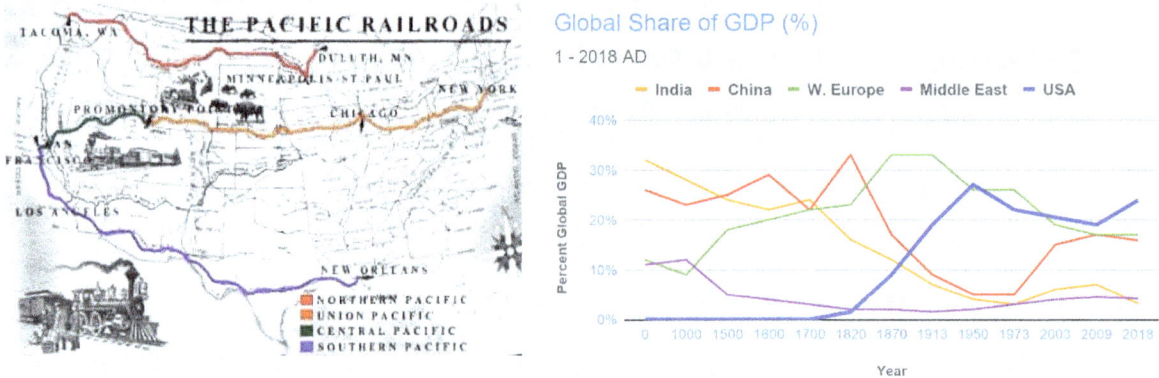

*The US was the first nation to build a transcontinental railroad contributing significantly to its rise as a great economic power in the 19th century (Image sources: Library of Congress/A. Maddison[34,35]).*

## US Space Logistics Infrastructure
### An Instrument of National Economic Power for the Future

On May 9th, 2019, a relatively small crowd gathered at the Washington DC Convention Center for the unveiling of Jeff Bezos' plans to build a commercial bridge to the Moon for delivery of cargo.[36] During the presentation Mr. Bezos explained that "infrastructure lets entrepreneurs do amazing things." Amazon, one of the most successful companies in US history, could not exist if it were not for existing infrastructure to include the internet, digital finance and overnight delivery. In order to posture the US space economy for accelerated growth, key infrastructure must be built upon which future businesses such as Amazon can flourish. The task is risky and expensive, but it has been successfully accomplished before. Key inflection points leading to previous unprecedented shifts in global economic and then military power were influenced by innovations in key infrastructure (see chart above). Two notable examples include the emergence of Western European maritime shipping during the spice trade in the 11th to 16th centuries, and the creation of the US transcontinental railroad accelerating the American industrial revolution in the 19th and 20th centuries. With regard to the latter, the US established itself as a great economic power in the 1870s, nearly seven decades before emerging as a great military power at the conclusion of World War II.

An early motivation for the transcontinental railroad was national security. President Abraham Lincoln recognized the need to protect the West from foreign claims and aggression. The railroad was built as a public-private partnership to achieve this objective and subsequently became the underlying infrastructure for interstate commerce and passenger travel. A great volume of commodities moved by rail to factories where products were produced and then to markets or ports for global distribution. Industries that were unimaginable at the time the railroad was first envisioned, flourished and contributed to a US great economic expansion. While we cannot fully imagine all of the industries that space will offer, a concerted US effort to be the first to establish logistics infrastructure in cislunar space is paramount to sustain future US economic leadership.

---

[34] Maddison, Angus. *Contours of the World Economy, 1-2030 AD.* Oxford:Oxford Press 2013. Print.
[35] World Bank. "World Development Indicators Database." 29 May 2020.
https://datacatalog.worldbank.org/dataset/world-development-indicators
[36] Blue Origin. "Going to Space to Benefit Earth (Full Event Replay)." YouTube video. 9 May 2019.
https://www.youtube.com/watch?v=GQ98hGUe6FM

- **Fractured open and free trade system** where knowledge, technology, talent and capital are restricted from flowing out of and into the US.
- **Russia and/or China become the space leader** providing lower-cost, higher capability space assets and services to US allies and partners.
- **A nation-state or non-state actor hacks existing space assets**, resulting in debris, technical glitches, and financial damage.

*The Apollo 16 crew captured this Earthrise during the second revolution of the Moon in April 1972. Much of the terrain seen here is never visible from the Earth, as the command module was passing over the far side of the Moon. (Source: NASA)*

## KEY ACTIONS & RECOMMENDATIONS

### SHORT-TERM

**Stand up a National Space Enterprise Task Force** - The Executive Branch establishes a National Space Enterprise Task Force that can evolve it into a permanent National Space Enterprise Center. The National Space Task Force would develop a "North Star" top-level vision and strategy to sustain US space superiority integrating and synchronizing actions across the whole-of-government, exercising all necessary instruments of national power. The National Space Enterprise Center, following a National Counterterrorism Center-like model, would serve as an interagency coordinating body, synthesizing diverse perspectives and rapidly resolving space-related issues across participating departments and agencies. (OPRs: NSpC, NSC, NEC, OTMP)

**Establish interagency export information sharing** - The US creates a common, export information system between the Departments of State and Commerce to ensure national security and global economic competitiveness. The NSC and NSpC s require all federal departments and agencies expand domestic information-sharing and tighten the export control review processes across the US government to speed up existing Departments of State and Commerce efforts to move technologies from the US Munitions List (USML) to the Commerce Control List (CCL). (OPRs: NSC, NSpC)

**Formalize information sharing with Five Eyes partners** - Pursue a Five Eyes-like collaborative agreement with allies and partners to develop a common standard, to formalize information sharing on space supply chain providers, and to support counterintelligence activities and training. (OPRs: DOC, DOS, DoD, ODNI, NASA, FBI)

**Establish new financial tools to increase transparency, liquidity, and investment capital in the new space commercial market** - Develop the range of short-term and long-term financial tools to support the rapidly evolving US space industrial base. These include tools to provide investment capital for companies, patient capital such as bond offerings for large-scale infrastructure, and exchange mechanisms that smooth future demand and supply curves. (OPRs: Treasury, DOC, DFC, SBA)

**Create a US-based Space Commodities Exchange** - The NSpC and the NEC should direct the Secretary of Commerce, the Chair of the Commodity Futures Trading Commission (CFTC), and other relevant experts to identify the necessary steps the US government must take to create a Space Commodities Exchange, and to report back to the NSpC on the way ahead. (OPRs: NSpC, NEC, OTMP, DOC, SBA, CFTC)

**Aggregate current and future US government market demand** - The Office of Science and Technology Policy (OSTP) funds a Federally Funded Research and Development Center (FFRDC) to conduct a federal inventory of existing and future US government demand for space-related goods and services and space-based commodities in the near-, mid-, and long-term. The US government should identify common purchases and develop a centralized process for government purchases of space-related products and services. (OPRs: NSpC, NSC, OSTP, OTMP, DoD, OMB, DOC, NASA)

# SPACE INFORMATION SERVICES

*"Truth will ultimately prevail where pain is taken to bring it to light."*

George Washington, President[37]

## BACKGROUND

**Decision making advantage** - In a world that is increasingly information driven, US civil, commercial and military power are ever more dependent on the quality of our information services. Superior information services are essential if the US is to outpace rivals and adversaries. They provide US advantage in the range and quality of data acquired, in the ability for uninterrupted and rapid transmission of data and information globally where, when and to whom required, in the ability for the data and information to drive faster and better decision making and in the ability for rapid decisions implementation.

**The increasing role of space information services** - Space systems are playing an increasing role in servicing national global information needs. This is being driven by the increasing diversity and speed of space information systems, technological advances that are transforming the capabilities and cost of these systems, increased demand and intrinsic global reach.

**Critical civil infrastructure** - Civil information services from space are essential for weather monitoring, weather forecasting, resource management, as well as climate monitoring, modeling and prediction. Space information services, such as the PNT provided by the GPS, provide essential resource mapping, city planning, and emergency response. Government information services enable broad economic sectors such as precision agriculture, banking, logistics tracking, and app-based taxi services. Specific space information services are essential to enable US leadership in space discovery and explorations to the Moon and across interplanetary space. Concerns at the highest levels of the US government over the manipulation and irresponsible use of GPS services are driving action to ensure responsible use and promote resilience.[38]

**Space information services enable the US economy** - Commercial space information services are a growing contributor to our economy. These include planetary-wide communication (satellite television, satellite radio, satellite data and global broadband internet) and active and passive Earth observation to monitor and forecast economic activity.

**Space information services convey military advantage** - In the defense realm, information services for global communications, overhead sensing, geolocation and timing are central to the success of US military operations across all the phases of conflict and across all domains (air, land, maritime, space, and cyber). They are central to indication and warning of our adversaries' capabilities, intent and action and drive every aspect of our find, fix, track, target, engage and assess (F2T2EA) "kill chain" and our ability to disrupt the "kill chain" of our adversaries. This enables faster and superior decision making

---

[37] Letter to Charles Mynn Thruston" Founders Online. 10 August 1794.
https://founders.archives.gov/documents/Washington/05-16-02-0376
[38] United States. 85 FR 9359-*Executive Order on Strengthening National Resilience Through Responsible Use of Positioning, Navigation, and Timing Services.* Washington D.C: Office of the Federal Register. 12 February 2020.

and economy in the application of force globally. As such, space information services are central to overall national defense and to the US ability to project space superiority and deny it to our adversaries over the phases of conflict as required for the cross-domain fight.

**Space information services shape global perceptions** - Space information services support the US informational component of power, enabling US public diplomacy to tell our story, shape the global narrative, and provide ground truth. Space information services provide ground truth to global audiences of human rights abuses, acts of aggression, and changes in military posture. They enable the US to communicate to otherwise denied areas. They enable the security of information for the US, allies and partners. Should the US lose the competition for space-based internet and Earth-based sensing, rival powers would have increased leverage to control the flow of information, conceal transgressions, and push propaganda.

**Hybrid space architectures to enable US military advantage and nurture a vibrant industrial base** - US military leadership recognizes that maintenance of information dominance requires transition from an eclectic mix of very capable but proprietary systems never intended to exchange information with each other to a scalable, hybrid space architecture of interconnected commercial, civil and military specific assets across a spectrum of satellite sizes, numbers, orbits and capabilities to optimize both mission performance and mission assurance. Doing so will require a shift to a zero trust, or variable trust, architecture that is focused on multi-factor authentication between users or nodes with the underlying philosophy of 'never trust, always verify.' Such a hybrid space architecture can leverage the accelerated growth in new space, cyber and emergent broadband communications capabilities while maintaining the integrity and security of space information services and the ability to project defensive and offensive, military space power.

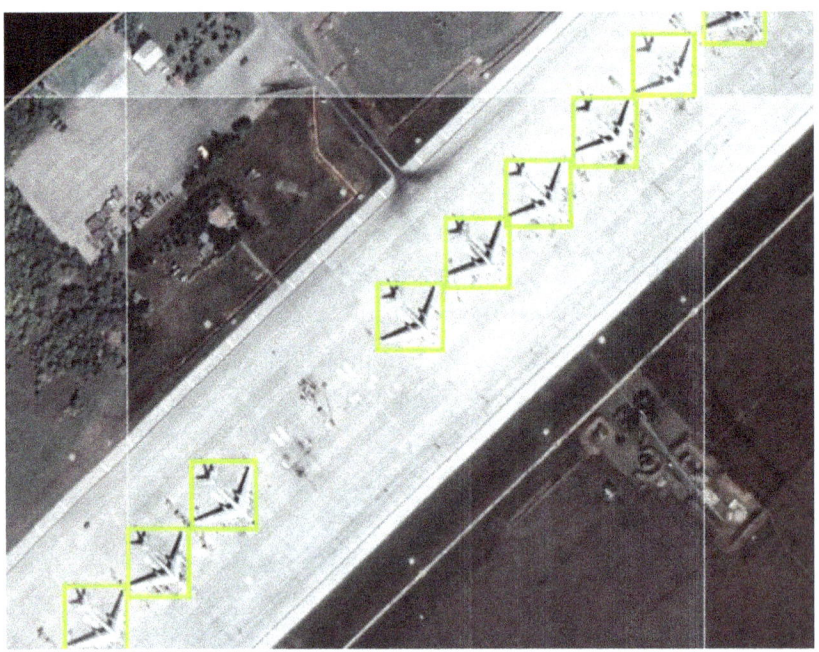

*Automated multi-class aircraft detection counted 53,974 aircraft in 1 day from 2,469 scenes and 342 areas of interest. It would normally take a trained analyst 7.5 work days on average to do the same. (Source: Orbital Insight/Airbus)*

**Common security needs** - Civil, commercial and military space information services have the common requirement to be secure against attacks from state or non-state actors who seek to deny, corrupt, or steal proprietary or national security information or assets, or seek to degrade the value and competitiveness of those capabilities.

## CURRENT STATE

Space information services are undergoing profound changes driven by increased demand coupled with decreases in launch costs, increases in information system performance and expansion of the range of platforms capable of providing those services. Space information services fall into several categories:

- **Space-based communication/internet and broadcasting** - Space communications/internet is currently dominated by geosynchronous capabilities provided by a range of national and international players SES (Luxembourg), IntelSat (Luxembourg/USA), EutelSat (France) and Telesat (Canada). This area, however, is undergoing a significant transformation towards a mixed architecture across a range of platforms and orbits most particularly low Earth orbit (LEO) systems to provide these capabilities driven by the decrease in satellite bus and payloads costs coupled with decreased launch costs. Emerging companies are working to deploy very-large constellations of small communications/internet satellites in LEO. These include OneWeb (UK, US), Space-X Starlink (US), TeleSat (Canada), and China Aerospace Science and Technology Corporation (CASC) Honyang constellation. In April 2020, China's National Development and Reform Commission (NDRC) added 'satellite internet' to a list of "new infrastructures" (also including 5G, Internet of Things and artificial intelligence) to be targeted for investment and guidance.[39]

- **Space Position, Navigation and Timing** - These capabilities are increasingly viewed, world-wide as part of a nation's critical national infrastructure resulting in a proliferation of national or regional space-based PNT systems (e.g. China/Baidu; EU/Galileo; US/GPS; UK, Japan, and India). This limits the international market for developing such national systems or providing this commercially.

- **Active and passive earth and space observing** - Presently, this area is dominated nationally and internationally by defense, civil and intelligence systems consisting of small constellations of primarily large satellites. Growth continues in the deployment of commercial constellations of smaller satellites to monitor and predict economic activity. As with communication/internet /broadcasting, cost, capability and demand changes are driving increasing exploration of a more hybrid architecture to meet defense, commercial and civil needs.

---

[39] Jones, Andrew "China's Commercial Satellite sector Sees Boost from 'New Infrastructure Policy." SpaceNews. 15 May 2020. https://spacenews.com/chinas-commercial-satellite-sector-sees-boost-from-new-infrastructure-policy/

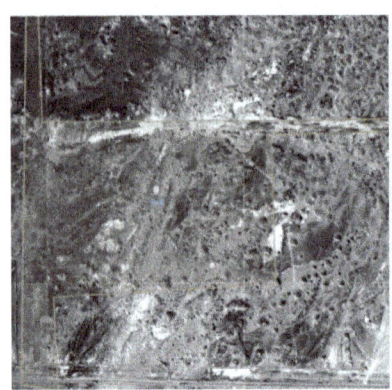  

2016 - No infrastructure     2017 - Construction     2018 - Expansion

*Imagery as ground truth showing construction changes year over year of Yuli Camp.. (Credit: Orbital Insight/Planet[40]).*

## "There is no such thing as re-education centers."[41]
### Exposing Ground Truth: Space as an Instrument of Information Power

Two years ago, reports first surfaced that hundred of thousands of China's Muslim Uighurs and other ethnic minorities were being involuntarily detained in 're-education' camps without trial. Authorities in Beijing initially denied the existence of these camps until human rights organizations and the international media provided indisputable evidence to show otherwise. Subsequent reports have revealed life behind these walled compounds - an environment where detainees are forced to renounce their religious beliefs and embrace the ideology of the Chinese Communist Party[42].

One of the greatest attributes of commercial GEOINT is that it is unclassified, which means it is in the public domain and therefore accessible to commercial customers, international partners and non-government organizations. It has been estimated that enough imagery will be collected in 2020 that 8.3 million human analysts, working 24-hours a day for an entire year, would be required to analyze the total collection. Detecting changes in a temporal dimension at countrywide scale (as illustrated in the photo sequence of Yuli Camp above) requires computer vision and machine learning algorithms leveraging the capabilities of cloud-based computing.

Space information services leveraging large constellations of small satellites and autonomous image processing is revolutionary. It is a product of US ingenuity, only made possible in the last several years, that will continue to improve over the next decade as commercial capabilities advance in spatial, spectral, and temporal resolution. As such, space information services contribute to national security as an important instrument of information power.

---

[40] Orbital Insight. "Using GO to Observe the Operational Status of Suspected Chinese "R-education" Camps." 16 September 2019. https://medium.com/from-the-macroscope/using-go-to-observe-the-operational-status-of-suspected-chinese-re-education-camps-717501ccf883

[41] Cumming-Bruce, Nick. qtd from CCP Official in "'No Such Thing': China Denies U.N. Reports of Uighur Detention Camps." New York Times. 13 August 2018. https://www.nytimes.com/2018/08/13/world/asia/china-xinjiang-un.html

[42] Buckley, C. and S.L. Myers. "China Said It Closed Muslim Detention Camps. There's Reason to Doubt That." New York Times. 9 August 2018. https://www.nytimes.com/2019/08/09/world/asia/china-xinjiang-muslim-detention.html

## KEY ISSUES & CHALLENGES

Space-based information services are transforming nationally (and even more so internationally) from large-satellite-centric architectures to hybrid architectures incorporating mixed capabilities, larger numbers of satellites, and satellites of diverse size and orbits, all with an increased capability to be networked in space. In this changing world the key issues and challenges are:

**Creating the needed market** - Ensuring a robust and growing, integrated US market across national civil, defense and commercial space information systems and services, to drive a vibrant and internationally competitive industrial base able to produce the new hybrid architectures. Incentivizing a vibrant industrial base by maximizing use of commercial capabilities and minimizing development of purpose built capabilities to meet civil and national defense needs.

**Secured capabilities** - Ensuring the development of a robust national capability for information services across civil, commercial and defense that is secure against threat or attack from state and non-state actors. Developing the required whole-of-government approach required to secure space information services as an essential element of national infrastructure.

**Global Competitiveness** - Ensuring US global competitive advantage in providing space information services in a fierce international market. Preventing competitors and adversaries from limiting the global information services market or disadvantage the US in that market by using anti-competitive practices to include government funding, theft of intellectual property, predatory control of supply chain, predatory investment in US firms or pricing, or state policies.

**COVID-19 impacts** - Ensuring the financial distress created by the COVID-19 pandemic does not severely damage companies that are driving the innovation that is necessary for the US space industrial base to meet and maintain leadership in space information services nationally and internationally.

## KEY INFLECTION POINTS

- **The commercial demonstration of technical and business viability of large constellations of small satellites** to provide communication, internet and or broadcast capabilities nationally and globally.
- **The demonstration of technical and cost viability of large constellations of satellites to meet major elements of defense requirements** for secure, global communication, active or passive Earth-observing, and/or civil needs for weather monitoring, resource management and environmental monitoring.
- **Demonstration of ability of hybrid commercial capabilities** to cost effectively meet the majority of defense or civil requirements in one or more areas of critical requirements, i.e. that commercial provides the majority of the capabilities required not a minor augmentation.
- **Demonstration of growth beyond niche applications of the market for commercial, Earth-observing capabilities** – convincing demonstration of a purely commercial growth path for space information services.
- **Success in minimizing effects of COVID-19 pandemic on range and innovation in the space industrial base for information services.**

## KEY ACTIONS & RECOMMENDATIONS

### SHORT-TERM

**The USSF institutes a "buy what we can; build only what we must" procurement approach** - Purchasing power should be used to promote a hybrid space architecture to include committing $1B/year on hybrid architecture through a commodities exchange and a diversity of sources. (OPRs: USSF, Congress)

**Increase investment by regulatory agencies in processes and framework** to keep up with the pace of innovation (Innovate by coming up with licensing to allow upgrades). (OPRs: DOT/FAA/AST, DOC/OSC, FCC)

### MID-TERM

**USSF specifies and commits to long-term architectural approaches and requirements clearly articulated to Industry** - In the mid-term (3-10) years, DoD commits to a hybrid space architecture. The DoD leverages the US commercial, small-satellite industry to develop and launch dedicated small-satellite systems that address major elements of requirements for mission assured tactical and strategic military capabilities. (OPR: USSF)

**Continue to lower multiple regulatory barriers and increase efficiency of regulatory processes for commercial enterprise** especially in regards to US interactions with allies. Revise export control laws for data products and development to remove historical barriers to international collaboration. (OPRs: DOS, DOC)

### LONG-TERM

**50% of government space information services are acquired from commercial sources.** (OPRs: DoD (USSF, DISA, NGA, NRO), USGS, NASA, NOAA, NSF)

# SPACE TRANSPORTATION & LOGISTICS

*"The US Military needs to focus on 'blue-water' space operations – GEO and above. US military space operations need to be in deep space, initially all of cislunar space, with an eye upon the entire inner solar system. To operate in deep space one needs to use the resources there, starting with fuel from asteroids. Once this is recognized, the military-economic imperative of identifying and protecting these assets becomes clear."*

Dr. S. "Pete" Worden, Brig Gen, USAF-Ret.

## BACKGROUND

Space transportation and logistics capabilities are critical for any expansive US space future across civil, commercial, and military domains. No nation has maintained a dominant position in a domain (air, land, maritime, space, and cyber) without a superior capability for movement and sustainment within that domain.

**The foundation of exploration** - In the civil domain space transport and logistics are foundational to US space exploration. They are critical in establishing a sustained human presence on the Moon as a stepping stone to Mars and beyond, for developing and maintaining the next cycle of great telescopes, and for the spacecraft and sensors that monitor the earth and explore the solar system. US capabilities to launch to space, maneuver in space, and sustain human and non-human space systems are essential to maintaining leadership in space exploration.

**The foundation for development** - In the commercial domain the ability to transport to and through space reliably and cheaply is the key determinant for the commercial viability of all other space capabilities. In the short-term this particularly drives the trajectory for personal space travel (space tourism, etc.) and supports the exploding demand for information services. In the mid- to long-term this drives the technical and fiscal feasibility of space manufacturing, resource extraction and space power.

**The foundation of military advantage** - In the military domain, space transportation and logistics will directly drive the ability to sustain superior space capabilities across the phases of conflict. These will determine the speed of inserting new systems, the number that can be cost effectively deployed, the ability of those capabilities to maneuver, to attack adversary capabilities as needed, to optimize the configurations of assets during combat and to repair and replenish assets. A robust space inspection capability, a byproduct of a logistics system, will allow early evaluation of anomalous events, attribution in the case of attack, and deterrence of aggressive activities.

**Multi-use logistical advantage** - A vibrant, robust space transportation and logistics industry can support a greatly expanded set of space activities, provide a clear US strategic advantage in space, and stimulate economic activity that is highly attractive to investors. Space transportation and logistics will become integrated into a single logistical system. Developing this system to support all users, civil, commercial and military, will be a powerful enhancement of national space growth and dominance.

## CURRENT STATE

**The US has regained its lead as the international leader in space launch** - In the last decade, US launch has dramatically increased from a single national provider to three or more providers for mid- to heavy-launch and several for small launch. SpaceX has demonstrated reusable launch systems, and Blue Origin and others are pursuing comparable capabilities. This has resulted in a rapid decrease in launch costs per pound across all launch categories. NASA and the Defense Advanced Research Projects Agency (DARPA) are pursuing nuclear thermal propulsion development for cislunar and interplanetary operations which will complement the space launch infrastructure by providing enhanced in-space mobility.

*Virgin Orbit's LauncherOne ignites in mid-air for the first time during the company's Launch Demo on May 25, 2020. (Source: Virgin Orbit/Greg Robinson)*

**The US is the global innovation leader in space logistics** - DARPA demonstrated in-space refueling as early as 2007. Earlier this year, the Mission Extension Vehicle-1, developed and funded by Orbital ATK (now a part of Northrop Grumman) successfully docked with a commercial satellite in GEO. This first-ever docking of two commercial spacecraft provides life extension services, a high-value economic activity. Additional new space logistics capabilities are in advanced development. NASA's Restore-L spacecraft will demonstrate refueling and in-space assembly, and will launch in late 2023. DARPA's Robotic Servicing of Geosynchronous Satellites (RSGS) program, being developed in partnership with Northrop Grumman, will provide repair and upgrade capabilities after its 2023 launch. Multiple commercial entities provide logistical support to the space station and NASA has recently awarded a similar contract for logistical support of sustained human presence on the Moon.

## KEY ISSUES & CHALLENGES

The primary challenges to US space transportation is maintaining continued growth in demand and the US market position in the face of increased competition in launch services. For in-space logistics, the challenge is to expand its application beyond space exploration into commercial and military operations.

**Maintaining US leadership in launch** - While the US has reestablished itself as world leader in space launch this lead is not assured. To the extent that nations determine that a launch capability is an element of critical infrastructure, those nations will develop space transportation capabilities with government funding and support, distorting the international market for space launch. Challenges are posed by non-competitive actions including theft of intellectual properties, investment in US start-ups to exploit their technology, and state subsidies of launch capabilities. In particular a major short-term challenge is the bankruptcy of small, start-up, US space launch and logistic companies and the acquisition of their intellectual and material assets by US rivals or adversaries.

**Increasing the market for launch** - Growth and advances in space transportation will be driven by demand. Recent decreases in US launch system costs have increased demand and allowed the US capture of an increased fraction of the global launch market. The challenge is to determine the actions that promote an ongoing virtuous cycle where decreased cost stimulates greater demand and greater demand stimulates economies of scale and innovation that decrease cost with the US as a dominant provider.

**Establishing the commercial and military role for logistics** - While the role and necessity for space logistics is well established for civil space exploration, the business case for commercial and the utility and cost case for military applications is still evolving. The challenge is to determine the role of logistics within a hybrid architecture for military and commercial capabilities where the hybrid architecture is a mix of small, relatively inexpensive satellites and carriers developed for frequent replacement, and larger and more complex systems that would benefit from refueling, repair, and upgrade supported by an in-space delivery system.

**Modularizing and commoditizing space systems** – A modularized and commoditized space paradigm would drive a transition from the chartered, payload-centric model of launch to a scheduled, cadenced launch model that lowers costs, increases efficiencies, and increases launch demand. The challenge is to demonstrate the feasibility of modularity and commodity distribution. This will expand the space industrial base beyond companies able to deliver fully integrated satellites to those that can produce space-qualified modules. This will accelerate introduction of new and innovative capabilities by requiring the development of a new module rather than an entire spacecraft. Similarly, reconstitution will become more feasible.

*(Left) Artist's concept of Space Logistics mission extension vehicle (MEV-1) approaching IntelSat's IS-901 satellite to provide life extension - a first for the commercial space industry. (Right) Photo of IS-901 taken by MEV-1 just before docking on February 25 of this year. (Source: Northrop Grumman)*

## First Commercial Satellite Servicing On-Orbit

Today, after a satellite is launched, it is completely on its own--never inspected, never maintained, never upgraded, and simply disposed of after it runs out of propellant. Satellites costing hundreds of millions or even billions of dollars are often completely operable when discarded. The company Space Logistics LLC, a subsidiary of Northrop Grumman, has taken the first concrete step to change that paradigm. For the last five years, they have developed the capability to extend the life of a satellite in geostationary Earth orbit (GEO). This is done by attaching a servicing satellite to a "client" satellite. Once attached, the servicing satellite's propulsion system is used to keep the "client" satellite in its proper position in GEO, allowing the client to continue generating revenue for several more years.

The life-extending spacecraft developed by Space Logistics is called the Mission Extension Vehicle (MEV). It accomplishes its mission by docking with the client's rocket nozzle, then pulling the two vehicles into a single joined stack. The first Mission Extension Vehicle (MEV-1) was launched in late 2019. After commissioning and checkout, it began a slow, careful approach to its client satellite, the IntelSat satellite IS-901, a nineteen-year-old satellite that provides communications to Europe and over the Atlantic Ocean. Successful docking of MEV-1 to IS-901 occurred on February 25th. Since then, MEV-1 has maneuvered the combined stack into position for IS-901 to continue providing communications services.

MEV-1 was completely privately financed. This event represents several notable firsts, including the first docking in GEO, the first all-commercial servicing event, and the first life extension of a satellite. Intelsat's CEO Steve Spengler has said, "If a satellite is still generating revenue, why would you want to throw it away?" Northrop Grumman continues to pursue even more advanced satellite servicing activities. They were recently chosen as the commercial partner for DARPA's Robotic Servicing of Geosynchronous Satellites (RSGS) program. RSGS will complement MEV by providing repair, upgrade and inspection services enabled by DARPA-developed robotic arms and tools.

## KEY INFLECTION POINTS

- **Return of human presence to the Moon** by the United States or a US rival or adversary driving space transportation and logistics technology and demand.
- **Successful demonstration of the commercial viability of large, small-sat, constellations for civil, commercial and/or military applications** driving increased development and launch of these systems, and removal of failed units to limit the increase of space debris.
- **Successful demonstration of the technical and commercial viability of sub-orbital space** driving the demand for on-orbit space tourism.
- **Successful demonstration of the commercial feasibility of logistically serviced space systems** to include modularity, repair, upgrades, and replenishment of commodities, and introduction of large space structures.
- **Successful demonstration of logistical servicing of a military system** that cost-effectively improves mission performance and/or mission assurance.
- **Commoditization of space services and material**, and development of robust space commodities markets.
- **New free-flying space laboratories** to augment and eventually replace the International Space Station, with increased use of robotics and a robust supporting logistics infrastructure.

*A proposed Unmanned Orbital Outpost leveraging a repurposed Dream Chaser 'Shooting Star' Cargo Module. Smart reutilization of space hardware enables experimentation and logistics while reducing costs and orbital debris. (Source: Sierra Nevada Corp)*

## KEY ACTIONS & RECOMMENDATIONS

### SHORT-TERM

**DoD initiates a program to incorporate dual-use space logistics into its space missions** (OPRs: USSF, AFRL, DIU)

**DoD commits to develop dual-use, civil/military, capabilities** for cislunar space domain awareness, establish cislunar awareness requirements, and develops a plan for a permanent presence at L1. (OPRs: USSF, DIU, AFRL, DARPA, SDA)

**Increase DoD/NASA collaboration and logistics infrastructure development** to include DoD engagement in Artemis, and developing a whole-of-government approach to lunar operations. (OPRs: OSD Space Policy, USSF, USSPACECOM, NASA, NSpC, NSC, NEC)

### MID-TERM

**Civil and military development of demonstrations of servicing and logistics** as elements of a hybrid architecture (OPRs: NASA, USSF, DARPA, DIU, AFRL)

**Military commitment to small launch** as an element of reconstitution of military assets. (OPRs: DIU, USSF)

**Resolve undetermined aspects of operation in cislunar space**, to include spectrum allocation, space traffic management, and debris management. (OPRs: NSpC, USSF, DOC)

### LONG-TERM

**Transition all GEO satellites and other high-value assets to a fully modularized architecture** with modules delivered by a cadenced launch regime to a reusable in-space transportation infrastructure. (OPRs: USSF)

**Integrate the DoD space architecture with a competitive commercial cislunar logistics market** in order to outpace adversaries, and in which commercial capabilities thrive without begging permission. (OPRs: DOC, NASA, USSF, NRO)

# HUMAN PRESENCE

*"The goal isn't just scientific exploration... it's also about extending the range of human habitat out from Earth into the solar system as we go forward in time... In the long run a single-planet species will not survive... There will be another mass-extinction event. If we humans want to survive for hundreds of thousands or millions of years, we must ultimately populate other planets.... I'm talking about that one day, I don't know when that day is, but there will be more human beings who live off the Earth than on it."*

Hon. Michael D. Griffin, Under Secretary of Defense for Research & Engineering[43]

## BACKGROUND

**The cusp of dramatic change** - The United States is on the cusp of a dramatic change in space access for its citizens driven by a revolution in the nation's spaceflight capabilities and in public and private sector attitudes on the 'how, who and why' of human presence in space.

**Changing public-private roles** - From its beginnings during the Cold War, human presence in space has been dominated by the US and other governments. Systems for human access to space were built for and controlled by governments. This is changing. While the government is still a prime investor in human space flight for exploration, government investment is now coupled with, and augmented by, ongoing and planned large-scale private sector investments. This trend is enhanced by a growing recognition on the part of lawmakers, NASA and DoD that a key element of a sustainable path for long-term US economic and strategic dominance is a catalyzed, vibrant and growing US private industrial base. The driving force for this private sector investment is both economic and aspirational with the goal of enabling human expansion into the Solar System and for space to become a domain of permanent occupation.

**Enabling & securing commercial human ventures** - The central finding for this area is that the US has an essential national interest in enabling and encouraging US citizens to venture into space, beginning with visits, and leading to the establishing long term presence and permanent human communities. Further, DoD in its role of securing space in support of national power must leverage these civil and commercial efforts wherever possible to enhance their mission and to provide a stabilizing and protective presence.

## CURRENT STATE

**The US is on the threshold of a new era in human presence in space** - Almost ten years after the retirement of the Space Shuttle, Space Exploration Technologies (SpaceX) launched on 30 May their DM-2 mission carrying two NASA astronauts to the International Space Station (ISS) for a 3 month mission. This flight is the culmination of the NASA Commercial Crew program for SpaceX's Dragon 2 and Boeing's Starliner transporting NASA astronauts to ISS and back. Equally important, this

---

[43] "NASA's Griffin: 'Humans Will Colonize the Solar System." The Washington Post. 25 September 2005. https://www.washingtonpost.com/wp-dyn/content/article/2005/09/23/AR2005092301691.html?noredirect=on

Commercial Crew Program is a major departure from traditional NASA practices with Boeing and SpaceX owning and operating their vehicles with NASA as a customer; hopefully one of many future customers. In the next several years a number of new, human-capable, launch systems are scheduled for operations including the ULA Vulcan/Sierra Nevada Dream Chaser, Blue Origin New Glenn and the SpaceX Super Heavy/Starship. They offer the promise of increasing launch-mass to orbit, to the Moon or Mars in excess of 100 metric tons and to reduce launch cost below $100/kg[44]. In addition both Blue Origin and Virgin Galactic are near commencing suborbital flights for personal space travel. Longer term both NASA and DARPA with its DRACO program are pursuing nuclear thermal propulsion.[45]

**Government - industry teaming is enabling lunar access and logistics** - NASA is aggressively pursuing the Artemis program to return humans to the Moon by 2024. Supporting that goal NASA is executing the Space Launch System/Orion heavy launcher program and has awarded contracts for the Human Landing System to take astronauts to and from the lunar surface with stays of up to two weeks. Awardees include: Dynetics, teamed with Sierra Nevada, Blue Origin teamed with Northrop Grumman, Lockheed and Draper, and to SpaceX with their lunar customized Starship.[46] NASA is developing the capabilities for a sustained presence on the Moon starting in 2028. NASA has also established the Commercial Lunar Payload Services (CLPS) program that will enable robotic commercial landers to help with advance survey and prepositioning, and to spur private access to the Moon.

**Government - industry teaming is enabling LEO industrialization** - Last year NASA unveiled a new ISS commercialization plan to encourage research and manufacturing and significantly, to allow two commercial astronaut visits to Station of up to 30 days each per year using Dragon 2 or Starliner. Also, Axiom Space won an award from NASA to use an ISS berthing node to host a commercial module offering commercial services with a 2024 launch date. Axiom recently announced that they had signed a contract with SpaceX to take 3 passengers (each paying $55 million) and an Axiom commander on a 10 day trip to the ISS in the second half of 2021. Space Adventures, which had previously flown 7 private astronauts to the ISS on the Russian Soyuz, has also contracted with SpaceX to perform a free-flyer mission that does not dock with ISS that could also send four private astronauts on a five day orbital flight in late 2021.

## ISSUES & CHALLENGES

**Expanding access** - While these technical advances are impressive the key issue is how to promote a virtuous cycle for commercial expansion of human presence in space beyond what is done for exploration and the fledgling capabilities for personal space travel offered by Virgin Galactic and Blue Origin. The challenge is to transition from personal space travel as a domain for the rich or a limited domain for civil human presence for exploration to one that supports multiple commercial activities in and through space and leads to a sustained presence of humans in space and on other celestial bodies and planets. This transition has three elements:

- Make it safer – Drive towards space travel being as safe as air travel is today.

---

[44] SpaceX.Starship: Overview. https://www.spacex.com. Web.
[45] Berger, Eric. "The US Military Is Getting Serious About Nuclear Thermal Propulsion." ArsTechnica. 15 June 2020. https://arstechnica.com/science/2020/06/the-us-military-is-getting-serious-about-nuclear-thermal-propulsion/
[46] NASA. "NASA Names Companies to Develop Human Landers for Artemis Moon Missions." 30 April 2020. https://www.nasa.gov/press-release/nasa-names-companies-to-develop-human-landers-for-artemis-moon-missions

- Make it desirable – Drive towards making travel to/from and habitation in space easy and enjoyable.
- Make it affordable – Drive down costs towards a price point <$100/kg that will enable millions of people to be able to afford to travel to space.

## KEY INFLECTION POINTS

- **Demonstration of technical and commercial feasibility of personal human transport to near space** by companies such as Virgin Galactic and Blue Origin.
- **Demonstration of the transport and logistics for human return to the Moon** and establishing a permanent lunar outpost, base or other sustained presence.
- **Success of routine US commercial, human, space-transport in cislunar space.**
- **Demonstration of a sustainable market for personal space travel.**
- **Demonstration of a technically- and economically-feasible commercial human habitat in space** as a destination for personal space travel.
- **Decrease in launch costs below $100/kilogram**; evidence of a Moore's Law for launch cost.
- **Demonstration of the benefits of and need for human presence in space for manufacturing or resource extraction.**

## KEY ACTIONS & RECOMMENDATIONS

### SHORT-TERM

**Sustain commitment to return human presence to the Moon by 2024** with an approach that contributes to advancing beyond exploration toward long-term human presence on the Moon. (OPRs: EOP, Congress, NSpC, NASA, DOC)

**Establish as a central national goal establishing a commercially, self-sustaining human presence in space** and a clear roadmap to achieve this end. (OPRs: EOP, NSpC)

### MID-TERM

**Commit to design and build a small-scale, in-space demonstration of a rotating habitat within 5 years** initially for exploration but with extension for commercial use. (OPRs: NASA)

**DoD specifies the role of human presence in space as a tool for military operations,** and DoD determines their role in the defense of US human occupied or operated exploration and commercial systems. (OPRs: USSF)

### LONG-TERM

**Demonstrate construction and operation of large, life supporting space structures** produced and assembled in space from in-space resources. (OPRs: NSpC, DARPA, NASA)

*NASA Astronauts Bob Behnken (left) and Doug Hurley (right) successfully execute the first commercial crew flight to the International Space Station aboard the US-flagged SpaceX Crew Dragon which launched from Kennedy Space Center on 29 May 2020. (Source: SpaceX).*

## Commercial Crew Program poised to save the USG up to $30B

Since the retirement of the Space Shuttle Atlantis nine years ago, the United States has not flown humans to space aboard a US space vehicle from US territory. That all changed on May 29th when NASA and SpaceX successfully executed the first commercial crew spaceflight in history. The significance of this mission cannot be overstated. Since the 1960's, NASA would identify a need for a crew transportation system and then the agency's engineers and specialists would oversee every development aspect of the spacecraft, support systems and operations plans. A commercial aerospace contractor would be chosen to build the system, ensuring that it meets the specifications spelled out by NASA. Personnel from NASA would be heavily involved and oversee the processing, testing, launching and operation of the crew system to ensure safety and reliability. All of the hardware and infrastructure would be owned by NASA.

NASA's Commercial Crew Program allows US companies to innovate and freely design in a way they believe is best, and they are encouraged to apply their most efficient and effective manufacturing and business operating techniques. The companies own and operate their own hardware and infrastructure. NASA's engineers and aerospace specialists work closely with the commercial companies, allowing for substantial insight into the development process and offering up expertise and available resources. The result is a commercial service that is scalable to address increased demand in civil, commercial and national security markets. The other commercial solution, Boeing *Starliner*, is expected to perform its next test flight later this year.

This government-private industry partnership has significant economic benefits, with more than 1,000 suppliers employing workers in all 50 states to support commercial crew spacecraft systems. While not done yet, the Commercial Crew Program is poised to save the US government approximately $20B-$30B, and provide two independent crew transportation systems.[47]

---

[47] McAlister, Phil. *Commercial Crew Program Status to NAC*. NASA. May 2020. https://www.nasa.gov/sites/default/files/atoms/files/ccp_status_to_nac_-_may_2020_1.pdf

# IN-SPACE POWER

> *"Clearly our first task is to use the material wealth of space to solve the urgent problems we now face on Earth: to bring the poverty-stricken segments of the world up to a decent living standard, without recourse to war or punitive action against those already in material comfort; to provide for a maturing civilization the basic energy vital to its survival."*
>
> Gerard K. O'Neill, Physicist & Visionary[48]

## BACKGROUND

**Space power is foundational** - The capabilities of all space systems now and into the future are critically dependent on the level of power either periodic or continuous that drive their operations. As such space power is foundational to define the range and effectiveness of a nations' space capabilities. Put succinctly, the US can't be a space power if the US doesn't lead in space power in both practical terms and as an aspirational element of overall space leadership.

**Space power enables exploration, presence, and mobility** - In the civil space domain, power is foundational to all exploration missions, and high power is a necessary capability to enable long-term human presence on planetary bodies or free space, to enable mobility and to develop industrial applications.

**Space power enables economic development and commerce** - In the commercial domain, space power is a foundational enabler for current services such as satellite communications and remote sensing, and the limiting factor for new markets in computation, commercial human habitats, in-space manufacturing, resource extraction, and space-to-space and space-to-Earth power beaming.

**Space power provides military advantage** - In the military domain, power enables peacetime surveillance, warnings and indicators, and is foundational to beyond line-of-sight command and control, secure communications, and precision targeting. Power advantage enables comparatively higher bandwidth, better connectivity, and broader applications. In conflict, an advantage in power means more energy to illuminate the conflict space, more energy to burn through jamming, and energy to enable maneuver.

## CURRENT STATE

Investigation of the current state of spacepower revealed the following major findings.

**Spacepower has become an arena of great power competition** - China and Russia publicly seek to challenge the US lead in space by seeking a quantum leap in space power, with announced roadmaps to build ambitious solar and nuclear space systems that will eclipse US capabilities.

---

[48] ONeill, Gerard K. *The High Frontier: Human Colonies in Space*. Burlington (Ontario): Apogee, 2000. Print.

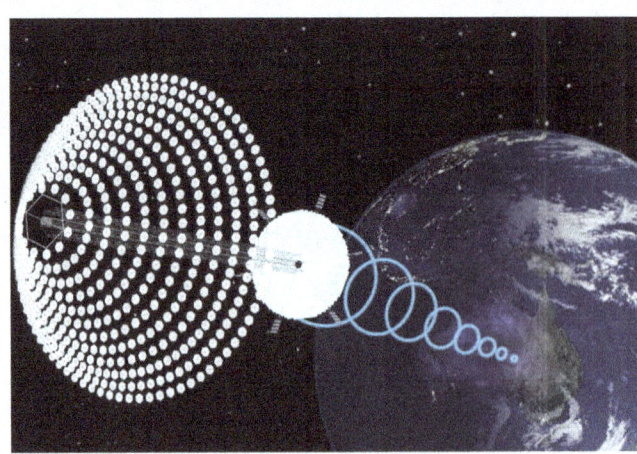

*The Solar Power Satellite via Arbitrarily Large Phased Array (SPS ALPHA) is a novel, bio-mimetic approach to the challenge of space solar power (Source: Mankins Space Technology).*

## China Pursues First Global Power Utility in Space

The US has been a leader in the study of space-based solar power technology since 1968 when NASA advisor and Apollo 11 project manager Peter Glaser first published his concept of a solar power satellite as a means of harnessing solar energy for transmission to Earth.[49] In 2008, the precursor to the USSF sought a mandate and funding to begin its development concluding that space-based solar power presents a "strategic opportunity that could significantly advance US and partner security, capability, and freedom of action and merits significant further attention on the part of both the US government and the private sector."[50] While a US program has failed to materialize, the People's Republic of China announced in 2010 a 40-year roadmap for space power. Executing this roadmap they are moving toward an initial operational capability for a high-powered solar power system by 2035. From its position of leadership in 2008, the US is now lagging and faces a situation similar to 5G, where inaction is enabling an aggressive adversary to potentially dominate a major element of the future space economy.

Space-based solar power has the potential to dwarf today's $400 billion satellite communications market. Moreso, it reduces dependency on fossil fuels and can be delivered to remote regions with little requirement for infrastructure. For these reasons, forfeiting US leadership risks ceding to China a dominant position in this global energy market. If China is successful, it puts at risk the US remaining an economic, military or political power in the second half of the 21st century.

The US is still positioned to compete and win in this area. The X-37B is flying a US Naval Research Laboratory experiment demonstrating the Photovoltaic Radio-frequency Antenna Module (PRAM) which converts solar power into RF microwave energy for transmission.[51] It is believed to be the first on-orbit space power experiment of this new and exciting technology.

---

[49] Glaser, Peter, "Power from the Sun: Its Future." Vol. 162, No. 3856. Science: AAAS. 22 November 1968. 857-861. http://www.science-sainte-rose.net/GrandBassin/documents/downloads/GlaserSPS68.pdf
[50] Office of Space Commerce. "NSSO Report on Space-Based Solar Power." 10 October 2007. https://www.space.commerce.gov/nsso-report-on-space-based-solar-power/.
[51] U.S Naval Research Laboratory. "First Test of Solar Power Satellite Hardware in Orbit." Phys.org. 18 May 2020. https://phys.org/news/2020-05-solar-power-satellite-hardware-orbit.html

**Power has become the limiting factor for many extant missions** - For decades, power was not the limiting factor and power investments could proceed at a low level of funding and effort. Recent innovations in launch, computation, packaging & deployment, lower cost of space hardware now mean that power is the limiting factor for many missions, including national security missions.

**Space power is foundational to major US space aspirations** - Power is a driver for satellite internet, low-Earth-orbit, human-tended and automated industrial facilities; in-space manufacture, lunar facilities, lunar and asteroid mining, and fast transit to Mars or the asteroids. These latter's requirements typically start in the megawatts.

**The US lead is not secure** - This industry is dynamic. The once steady market of geostationary communications satellites has largely collapsed, with new demand coming from a more sporadic growth of Low-Earth-Orbit mega-constellations. Today, two US space solar cell vendors, SolAero and SpectroLab, command the largest global market share, but the combination of state-supported competition from China and Europe and insufficient incentives to 'buy American' put both at risk. In particular, the state of US investments may allow competitors to fully automate their production lines (which could halve costs), and then dump products into the market, putting American suppliers out of business.

**The US lacks a coordinated and consolidated national vision and policy for space power** - Unlike the clarity of the PRC's roadmaps or our own Moon 2024, there is no US national vision and policy which advances US power toward clear, time-specified goals, or ties space power accomplishments toward a broader space industrial strategy and vision.

*The Photovoltaic Radio-frequency Antenna Module (PRAM) with a 12-inch ruler for scale. The hardware is the first orbital experiment designed to convert sunlight for microwave power transmission for solar power satellites. (Source: US Naval Research Laboratory).*

## KEY ISSUES & CHALLENGES

**Great power competition** - Organized competitors are seeking to unseat the US as the global space power leader. They are working to beat America in reaching critical, solar and nuclear, high-power tipping points that open new 'bootstrapping' self-sustaining markets and expanded freedoms of action. Short-term, the US may lose its position as the global provider for the mega constellations while in the mid-term, the US may suffer a hit to prestige when rival powers orbit larger power systems. Most consequentially, in the long-term the US lagging in space power development will pervasively impact overall national power in space, with the US losing out on resource extraction, in-space manufacture, and utility-scale power markets. For national defense that could translate into a loss of the positional, industrial and logistical advantages affecting both the mass and means for projecting military power in space

**Consistent and adequate funding over time** - Without significant investment, the US is likely to lose its position of global supplier for satellite power systems. The US must maintain a long-term commitment to advancing space power development through systematic government investment in technologies over significant time periods that cross administrations. Historically, innovation in the power sector has been led by government investment in basic research, funding first implementation in commercial systems, followed by use in government systems.

**A need for government leadership** - Power, as a fundamental enabler, is among the most appropriate technologies for a 'technology push.' Historical requirements-driven technology pull efforts have driven only incremental advances. While the private sector would respond to incentives, those incentives currently do not exist. Significant regulatory risks exist for new applications such as power beaming, which would require the security of an active partner in the US government. The significant technical risks and uncertainties of private companies could be greatly reduced with the security of a government customer.

**A national vision for space power** - There is no coordinated, consolidated national vision and policy for US space power providing a strong demand signal to US research and development agencies and the private sector. The US lacks the clarity of either its Moon 2024 or the PRC's roadmaps to 2050 to provide clear, time-specified goals which tie space power milestones to a broader long-term competitive industrial strategy and vision. Such a vision and associated policy can establish the US government as an anchor or lighthouse tenant of early, space-power, capabilities and of lunar fuel production enabled by space power to drive development of capabilities which enable broader and self-sustaining commercial markets.

## KEY INFLECTION POINTS

- **China emerges as the primary photovoltaic supplier** for the mega-constellations.
- **Development of power systems to support human presence in Low Earth Orbit (LEO)** and at Lunar Poles which drive substantial power capabilities, and enable broader 'bootstrapping' industrial activities.
- **Launch of the first space reactor after 50 years of non-activity** by the US or rival power reigniting competition and progress in space nuclear power and propulsion.
- **In-space assembly of large area platforms and off-earth manufacture of power system components** to enable very-large in-space manufactured power systems.

## KEY ACTIONS & RECOMMENDATIONS

### SHORT-TERM

**Establish a comprehensive, national space power vision** equivalent to those of our rivals and competitors. (i.e. it should include equivalent ambitious industrial goals and an equivalent timeline to 2050, and provide specific milestone dates). This vision should be linked to an overall "North Star" space industrial vision and policy and specific Executive Order tasking agencies and departments. (OPRs: EoP, NSC, NSpC)

**Fully fund existing programs to maintain parity in great power competition** - This includes the following programs: AF SSPIDR, DOE/NASA Kilopower & NTR, DARPA DRACO, OSD/NRL OECIF SBSP/Powerbeaming Roadmap]. (OPRs: Congress, OUSD(R&E), USSF, NASA, DOE, DARPA, OSD, NRL)

**Develop deep-space CONOPS and requirements** identifying advanced, space power and power beaming needs, and establishes public-private partnerships and anchor tenant agreements to secure logistical advantage. (OPRs: USSF, DIU)

## MID-TERM

**Expand the AF Space Solar Power Incremental Development and Research Program (SSPIDR)** to include other system concepts and component tech to provide adequate technical trade space for operational, high-power space solar power systems and ensure the US is first to break the meter-squared threshold. (OPRs: AFRL, OUSD(R&E), Congress)

**Fly demonstrations of 1, 10, and 100kw nuclear systems.** (OPRs: NASA, DOE, DARPA)

## LONG-TERM

**Broker anchor tenant agreements for early commercial space power capabilities and propellant production on the Moon** enabled by space power for US agencies. (OPRs: DOC, NASA, USSF)

**Use requirements of National Near Earth Object Preparedness Strategy and Action Plan to drive advanced power and propulsion** to prevail in the contest for dual-use deep space movement and maneuver. (OPRs: USSF, USSPACECOM, DOE, NASA, DARPA)

*This page left intentionally blank.*

# SPACE MANUFACTURING & RESOURCE EXTRACTION

*"Our objective in returning to the Moon is to learn how to live and work productively on another world. The Moon possesses the material and energy resources necessary to learn new skills to create new space faring capabilities. Its proximity to the Earth permits easy and routine access to its surface for just such an endeavor that, if successful, will serve as the catalyst and the true historical starting point for human expansion off-planet."*

Paul Spudis, Planetary Scientist[52]

## BACKGROUND

Space manufacturing includes products manufactured in space for both terrestrial and space economies, and products that support in-space industrial infrastructure enabling the expansion of the commercial space market. Space resource extraction encompasses the removal of valuable minerals and water from the lunar surface and asteroids. Resource extraction from space is critical to enabling the space manufacturing market and in establishing space outposts, habitats, power sources, and propulsion. Common space resources such as iron, aluminum, and titanium are essential to electrical and structural components, while silicon is a raw material for solar panels and computers. Extracted water, while essential to sustain life, can also be broken down into hydrogen and oxygen, to meet a variety of needs – oxygen is breathable, recombining hydrogen and oxygen generates electrical power, and liquid hydrogen and liquid oxygen can serve as a propellant. A conceptual illustration of the space manufacturing and resource extraction process is shown in Figure 1.

## CURRENT STATE

**National Purpose** - In 2015, the US Congress passed Public Law 114-90, which established the legal right of Americans to engage in commercial recovery of space resources and to own, possess, transport, use and sell such resources and for the President to facilitate the exploration for and commercial use of space resources. US policy states, "Americans should have the right to engage in commercial exploration, recovery, and use of resources in outer space, consistent with applicable law. Outer space is a legally and physically unique domain of human activity, and the United States does not view it as a global commons. Accordingly, it shall be the policy of the United States to encourage international support for the public and private recovery and use of resources in outer space, consistent with applicable law."[53] To encourage such support, NASA has introduced the Artemis Accord principles[54]

---

[52] Spudis, P. & Lavoie, A. (2011, September). Using the resources of the Moon to create a permanent, cislunar space fairing system. In AIAA Space 2011 Conference & Exposition (p. 7185).
[53] United States. 85 FR 20381. *Executive Order on Encouraging International Support for the Recovery and Use of Space Resources*. Washington D.C: Office of the Federal Register online via Government Publication Office. 10 April 2020. https://www.govinfo.gov/content/pkg/FR-2020-04-10/html/2020-07800.htm
[54] NASA. "Principles for a Safe, Peaceful, and Prosperous Future." 2020. https://www.nasa.gov/specials/artemis-accords/index.html

as a condition of participation in the Artemis Lunar project to shape the conditions through individual bilateral agreements.

**The US is leading** - The US is currently the technological lead in space manufacturing and space resource extraction. During the shuttle era, the US explored numerous space manufacturing technologies including structural assembly, beam-building, and the 'wake-shield' experiments in-space manufacture of integrated circuits and photovoltaics. More recently, the ISS has hosted US firms experimenting with products that can only be manufactured in microgravity: specialty optical fibers, 3-D printed organs, and corneas. NASA's Solar Power satellite programs in the 70s and 90s significantly advanced concepts for in-space manufacture and resource extraction. NASA has helped mature resource extraction technology through its 'Swampworks,' Centennial Challenges, and NIAC grants. While resource extraction is yet to be proven, several US startups are pursuing lunar and asteroid resource extraction.

**Figure 1.** Conceptual space resource extraction and manufacturing process.

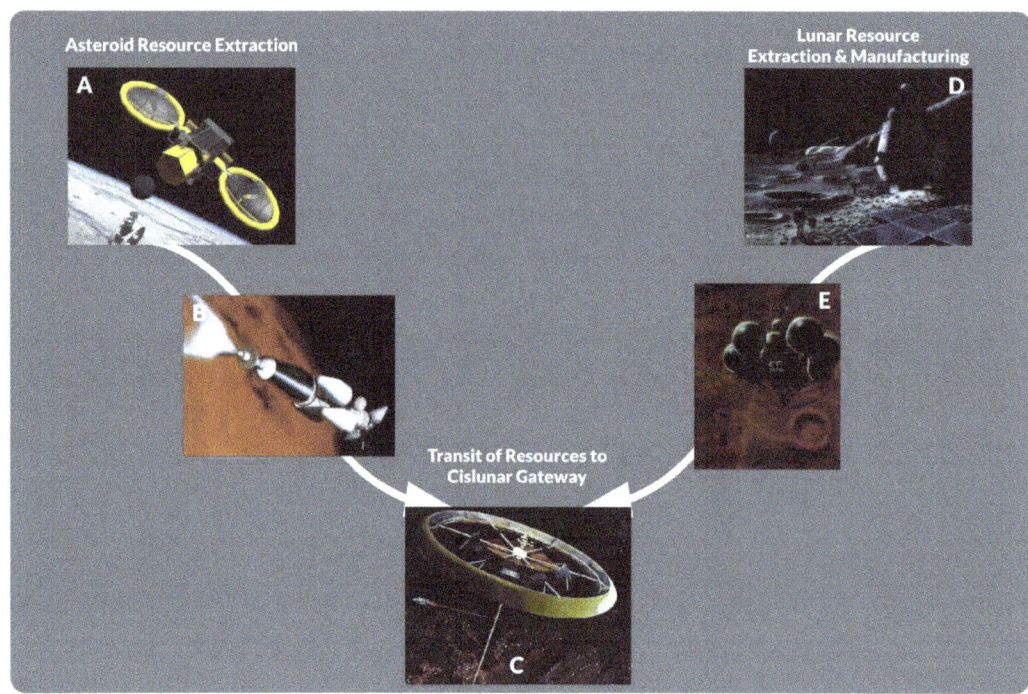

*Asteroids are captured and mined for resources by robotic spacecraft (A). Resources from this asteroid mining process are transported (B) to an on-orbit manufacturing facility and cislunar commercial gateway (C). Minerals and water are harvested on the lunar surface and used to manufacture lunar structures and generate power (D). Excess lunar resources are transferred from the lunar surface (E) to the cislunar commercial gateway (C) [Image sources: (A) Artist concept of NASA's OSIRIS-Rex: NASA/Goddard/Chris Meaney, (B-E) NASA Advanced Space Exploration Project Art.]*

**Great power advantage** - The combination of space resources extraction and in-space manufacturing provide great power advantage in the wealth they can generate and in the advantage they can provide and in the range, mass, flexibility and cost of a nation's space capabilities available to exert its power and influence in space and beyond. While most applications of space manufacturing and resource extraction will be realized in the mid- to long-term, they are foundational to enable cost efficient, expanded, and robust capabilities in space, to include computing, sensing, maneuver, unique platforms, large structures, fuel, and power. Ceding space manufacturing and resource extraction advances to other state actors could tempt industry to migrate to the more vibrant power.

*In-space manufacturing and assembly of large spacecraft and habitats will not be practical without first harnessing the tremendous resources of the Moon and other near Earth objects: (Source: SkyCorp).*

## "Whoever first conquers the Moon will benefit first."
### The strategic importance of lunar in-situ resource utilization

The Earth is the most massive rocky world in the inner solar system. This mass comes at a price - launch is expensive from the Earth's surface into space. For example, a direct trip from the Earth to the surface of Mars would exhaust half its total fuel simply to reach a 400 km orbit about the Earth. Our planet sits at the bottom of a deep gravity well, whereas the Moon does not. Therefore, lunar resources provide a tremendous competitive advantage to the first industrial power that learns to utilize them.

What kinds of resources? The Apollo missions showed that the Moon's surface composition includes significant amounts of elemental oxygen, silicon, iron, aluminum, calcium and magnesium. Subsequent NASA missions have identified deposits of titanium ore up to ten times larger than those found on Earth, and water ice deep within permanently shaded 'cold traps' in lunar craters at the poles, potentially as much as 600 million metric tons. This water ice can be used for potable water, oxygen to breathe and rocket fuel (liquid hydrogen and oxygen). Many other metallic species may reside within the Moon's many impact craters.

Today's race to the Moon has little to do with flags and footprints. Strategically, it is a race to the great wealth of lunar resources which will fuel the greater space economy and enable future exploration and settlement in the solar system. Ouyang Ziyuan, Chief Scientist of China's lunar exploration program, recognized this in 2002 when he stated that *"whoever first conquers the Moon will benefit first."*[55] To date, no volumetric mineral estimate of the Moon has been accomplished. This is a high priority for remote sensing and lunar exploration missions as well as the technology development required to harness the vast resources of the Moon. US forfeiting access to lunar resources risks perpetuating the kind of bind it is experiencing today with limited access to rare Earth metals which are 90% controlled by China.[56]

---

[55] Goswami, Nasrata. "China's Get-Rich Space Program." The Diplomat. February 28 2019. https://thediplomat.com/2019/02/chinas-get-rich-space-program/

[56] Woodward, Aylin. "China could restrict its Export of Rare-Earth metals as a Trade-war tactic." BusinessInsider. 4 June 2019. https://www.businessinsider.com/rare-earth-metals-elements-what-they-are-2019-6

**Low private investment** - Currently, there is limited private investment in space manufacturing and resource extraction due to limited demand for space products and technological barriers to space manufacture. However, NASA and, to a lesser extent, the DoD, have invested in this area. Through public-private partnerships, NASA is working with TransAstra to fund space demonstrations of asteroid resource extraction through optical mining.[57] TransAstra through Small Business Innovation Research (SBIR) funding is exploring prototypes for a lunar polar propellant mining outpost.[58] Lunar Resources with DoD SBIR funding is pursuing on-orbit manufacturing of a large mirror for hypersonic missile tracking.[59] Lunar Resources has also put forward proposals to extract minerals from the lunar regolith and manufacture large solar farms directly on the Moon. Made in Space is pursuing on-orbit manufacturing programs, to include additive manufacturing in space and Archinaut, a space platform able to assemble spacecraft components in low-earth orbit.[60] Multiple commercial companies are pursuing in-space assembly, refueling, and the manufacture of bio products. The DoD is advancing foundational technologies through SMC's Advanced Space Testbed, DIU's Outpost and Multi-Orbit Logistics, and AFRL's Cislunar efforts.

*Robotic construction of landing pads, aprons, roads, blast walls, shade walls, thermal and micrometeorite protection shields and dust-free platforms is being developed based on ISRU-based resources. (Source: NASA)*

**Critical dual-use enablers** - Total US investment in this area is estimated to be in the range of $100 - $200 million, with most of that funding currently focused towards on-orbit manufacturing in Earth orbit.[61] Of note, space transportation and logistics and space power are key enabling areas critical to the expansion of space manufacturing and resource extraction across the cislunar domain. Identifying, funding, and demonstrating dual-use applications (e.g. near-term DoD application and long-term commercial application) are key to developing this area.

## KEY ISSUES & CHALLENGES

**Communicating the long-term benefit** - The greatest challenge for this area is communicating the long-term civil, commercial and national defense benefits of overall space infrastructure (e.g. power, outposts, etc.) and space manufacturing and resource extraction. Similar to past significant infrastructure investments, such as the interstate highway system and the internet, space infrastructure

---

[57] Sercel, Joel "Mini Bee Prototype to Demonstrate the Apis Mission Architecture and Optical Mining Technology." TransAstra Corporation. 11 June 2019.
https://www.nasa.gov/directorates/spacetech/niac/2019_Phase_I_Phase_II/Mini_Bee_Prototype/
[58] Sercel, Joel "Lunar Polar Propellant Mining Outpost (LPMO)" NASA. 7 April 2020.
https://www.nasa.gov/directorates/spacetech/niac/2020_Phase_I_Phase_II/Lunar_Polar_Propellant_Mining_Outpost/
[59] Erwin, Sandra "Space Startup Developing Technology to Detect Hypersonic Missiles" SpaceNews. 19 December 2019.
https://spacenews.com/space-startup-developing-technology-to-detect-hypersonic-missiles/
[60] Harbaugh, J. OSAM-2, 23 April 2020, nasa.gov.
[61] Harbaugh, J. OSAM-2, 23 April 2020, nasa.gov.

has the potential to yield dividends economically and politically for decades after establishment. As in many grand infrastructure projects, government investment is critical to seeding the effort and buying down risk for private industry.

**Launch cost, development time and demonstration opportunities** - Other key challenges to space manufacturing and resource extraction are: 1) reducing launch costs; 2) reducing time from concept to launch, and 3) incentivized demonstrations related to this area.

**Robotics** - Due to the remote and harsh environment, leveraging improvements in trusted autonomy and robotics are critical to harvest space resources and to use them to generate products in space.

**Establishing international norms, standards and law** - There is a critical need to articulate international norms, standards, and laws related to space resource extraction and manufacturing, such as the Artemis Accords, as well as protecting intellectual and real property terrestrially and in space are critical to successful progress in this area.

## KEY INFLECTION POINTS

- **Initial and sustained USG investment** to seed development of this long-term area. (Improved space experimentation and demonstration processes could accelerate investment in this area.)
- **Reduced launch times, expedient licensing, improved return dynamics** to increase the access to space to demonstrate the utility of space resource harvesting and manufacture.
- **Demonstration of in-space construction of large structures** by the US or rival power.
- **Emplacement of sufficient power at the lunar poles** by the US or peer competitor to enable industrial-scale activities on the Moon including mining.
- **Demonstration of harvesting water on the Moon** for usable propellant by the US or rival power.
- **Demonstration of extraction and processing of minerals on the Moon** by the US or rival power.
- **Demonstration of asteroid mining** by the US or rival power.
- **The first commercial sale of a space resource**, including to a government buyer.

## KEY ACTIONS & RECOMMENDATIONS

### SHORT-TERM

**USSF articulates a vision and requirements for space resources extraction and capabilities they enable.** (OPRs: USSF, USSPACECOM)

**The US government provides a reasonable level of initial and continuing funding** to seed this area, buy down risk, and encourage venture capital investment. (OPRs: NASA, USSF/CSRO, DOC, DIU)

**DoD advances dual-use logistics solutions** through broad-reach, flexible funding mechanisms along the lines of DARPA's Grand Challenge or AFWERX challenges. (OPRs: AFRL/RV, DIU, SDA, DARPA)

**US government space agencies sponsor space manufacturing and planetary defense chairs and programs at universities** to build basic research capacity and the workforce (OPRs: USSF, NASA)

## MID-TERM

**The US government civil and national security space agencies serve as an "anchor tenant"** by creating market demand for propellant from space. (OPRs: USSF, NASA, DOC, DIU)

**The US government develops a national strategy that markets the cislunar economy** to signal US interest to industry and academia and provides time-specific milestones similar to China's long-term vision. (OPRs: NSpC, DOC, NASA, USSF)

## LONG-TERM

**The US government takes leadership to create a public-private lunar industrial park** conceived from the beginning to enable commercial partners to scale to high volume, high capacity systems. In contrast to ISS which is not set up to do more than tech demos, the outpost architecture must include links to support resource extraction and on-orbit manufacture. (OPRs: NASA, DOC)

**The US government leads implementation of large-scale cislunar projects** to spur continued development of industrial capability (OPRs: NASA, USSF, DOC)

# INSIGHTS & RECOMMENDATIONS FOR INDUSTRY

## A TWO-WAY STREET

While this report's primary aim is to provide actionable recommendations to the US government to ignite and enable a powerful US industrial base, the successful implementation of these recommendations requires an active role and partnering with industry. Because of this inter-dependence a number of parallel key actions and recommendations for industry were detertermine in the course of the workshop and in the development of this report.

*Orion is the first spacecraft designed for long-duration, human-rated deep space exploration. It will transport humans beyond low Earth orbit to destinations such as the Moon and then return them safely back (Credit: Lockheed Martin).*

## KEY ACTIONS & RECOMMENDATIONS

**Identify opportunities for public private partnerships** - Industry should be proactive in proposing and entering into public-private partnerships to develop dual-use, commercial/government space capabilities and enabling technologies. These partnerships should be joint funded. They should focus on developing capability whose commercial viability benefits from but is not strongly reliant on income from the US government. Such action on the part of industry helped shape NASA's successful Commercial Orbital Payload Service and the Commercial Crew Program, and can similarly shape DoD and USSF thinking and actions. Industry insights can help shape future architectures, concepts of operation, and requirements. The DoD now has a breadth of opportunities for 'Space Act-like' public private partnerships using negotiable Other Transaction agreements.

**Industry must proactively protect itself from predatory exploitation** - Our adversaries know that many of the most disruptive ideas driving the future of space come from the US industrial base. Our adversaries, particularly China, have an ongoing and extensive program to acquire those ideas by any means possible. US industry must do better in protecting this valuable intellectual capital. Entrepreneurs can contact their local Federal and Department of Defense Law Enforcement and Intelligence Community members for more information on reducing vulnerabilities related to economic espionage and concerns of intellectual property and trade secrets theft.

**Guide and develop the future STEM workforce** - Talent, vision and education will drive the future of the US space industrial base. Our educational system is not structured or incentivized to deliver the STEM workforce to drive the space industrial base to create the future we all want. While government has a role, industry must be much more proactive, independently and in partnership with government, in working across the US education system to develop the needed STEM talent. This means engagement in K through 12, endowing university chairs in areas directly supporting industry personnel needs, funding undergraduate scholarships/loans for STEM students, internships and providing space professionals to support instruction in space subjects.

**Pursue supply chain hygiene** - Secure supply chains are critical to commercial success and the ability to provide services and capabilities to the US government. Industry needs to increase their understanding of the commercial and security vulnerabilities within their supply chains and take proactive steps to counter those vulnerabilities. Part of their strategy to address these vulnerabilities is for industry to improve ties and partnerships with domestic and allied, parts, subcomponent and subsystem manufacturers to strengthen trust and resilience in space supply chains. Wherever possible, industry should choose domestic or allied manufacturers of parts and subcomponents.

# SUMMARY & CONCLUSIONS

> *"The exploration of space will go ahead, whether we join in it or not, and it is one of the great adventures of all time, and no nation which expects to be the leader of other nations can expect to stay behind in this race for space."*
>
> John F. Kennedy, President (1962)[62]

## FINDINGS

The world is entering a new and exciting era for space. In this century space will continue to grow rapidly as a major element of overall human civil, commercial and military actions and as an element of any country's national power. This new era offers promise and hazards to the United States. Continued leadership in space will anchor US national power. Loss of leadership will put the US global strategic interests at risk. Key to any future with the US as a space leader requires a vibrant, innovative robust US space industrial base.

The US is a great space power and is positioned to retain space leadership if the right strategy and actions are developed and implemented. The monumental achievement by NASA partnering with US entrepreneurs and industry, SpaceX has returned human spaceflight to US soil. Behind the scenes is a strategic US space sector in transition where innovative partnerships between the US government and commercial industry have the potential to accelerate this new frontier economy and provide the next strategic US offset. Trillions of dollars of human economic activity are moving into Earth orbit and beyond. US leadership can ensure that the millions of jobs that will be created will be American, and that the US remains the world's leading economic superpower through the 21st century

This report documents the results of a major conference and workshop that gathered national space thought leaders and key participants across the US space community. The conference analyzed in detail the overarching trends and challenges facing the industrial base and US space power. The conference successfully executed the next higher level analysis of the challenges to US space power and the required action to sustain US space power and the US space industrial base to support that power. The conference results go beyond "defining the problem" to providing initial, executable recommendations on the actions needed to secure a future of US space leadership. Six overarching recommendations for government and four for industry were synthesized. These are presented in more detail in the above report and summarized in the executive summary. In addition, specific recommendations were generated for each of the six more detailed areas of or affecting the space industrial base.

This analysis and recommendations provide a strong basis for immediate action and form a solid foundation for continued development of an integrated comprehensive national space vision, strategy and execution plan. The organizers and participants in the conference look forward to this report being the basis for action and continued vigorous discussion.

---

[62] "Text of President John F. Kennedy's Moon Speech at Rice University." Ricetalk. 12 September 1962. https://er.jsc.nasa.gov/seh/ricetalk.htm

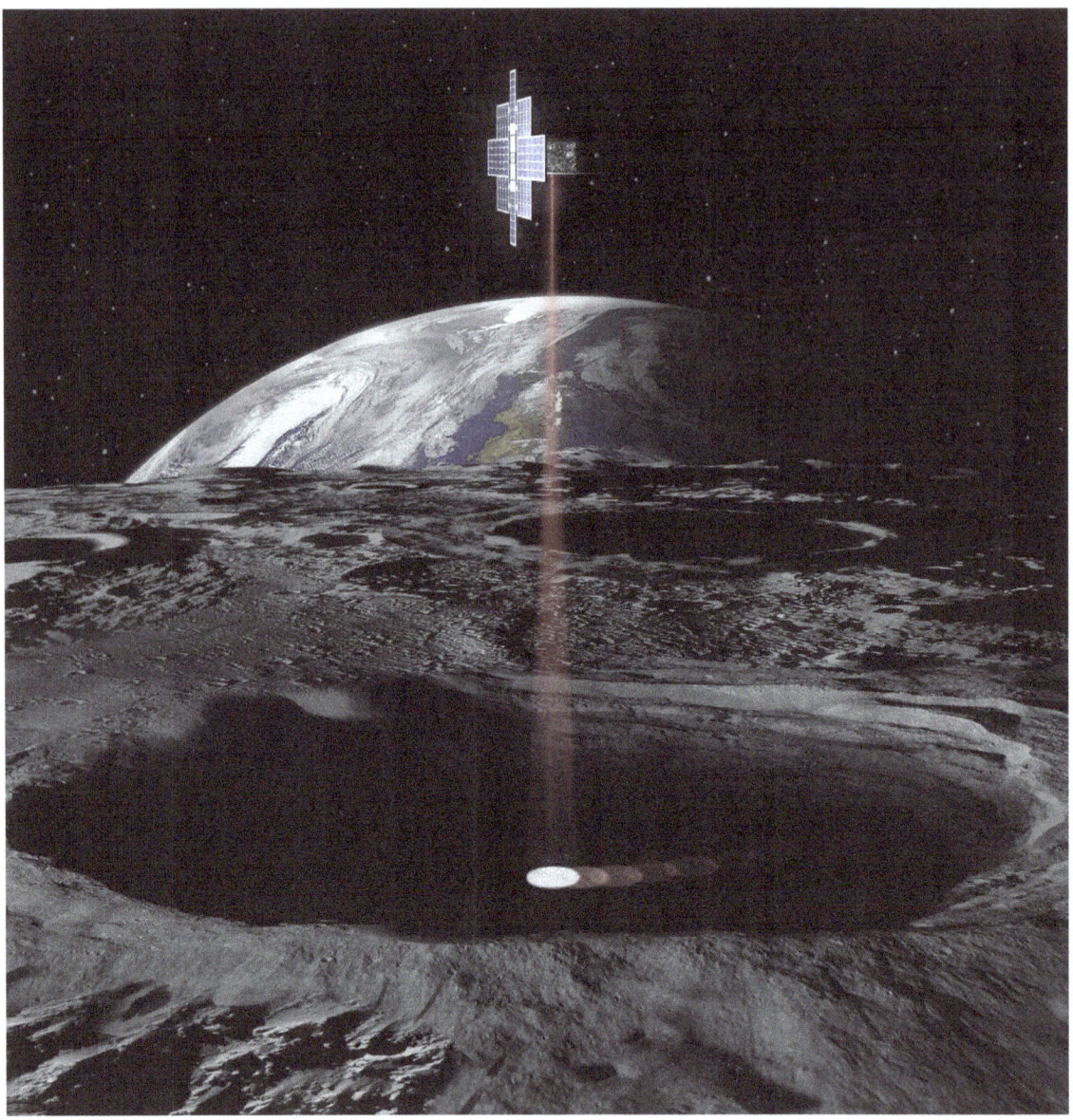

*The Lunar Flashlight spacecraft, a six-unit CubeSat designed to search for ice on the Moon's surface using special lasers. The spacecraft will use its near-infrared lasers to shine light into shaded polar regions on the Moon, while an onboard reflectometer will measure surface reflection and composition. (Source: NASA/JPL-Caltech)*

# APPENDIX A
## PARTICIPANTS

Lt Col Raj Agrawal, Dept of the Air Force
Brett Alexander, Blue Origin
Albert Arquer, Capella Space
Jason Aspiotis, Finsophy PBC
Martin Attiq, Astra
Jatin Bains, Space-Eyes
Payam Banazadeh, Capella Space
Dr. Jeremy Banik, NRO CSPO
Avram Bar-Cohen, Raytheon
Tammer Barkouki, Defense Innovation Unit & UC Davis NASA HOME Space Tech Res. Inst.
David Barnhart, Arkisys Inc.
Klay Bendle, Defense Innovation Unit
Scott Bennett, NewSpace Consultant
Brad Blair, NewSpace Analytics, LLC
Michael Bloxton, Bloxton Investment Gp, LLC
CDR Alan Brechbill, US Space Force
Brig Gen Steve Butow, Defense Innovation Unit
Bruce Cahan, Urban Logic, Inc.
Elliot Carol, Lunar Resources, Inc.
Dr. Daniel Ceperley, LeoLabs
Dr. Brad Clevenger, SolAero Technologies
Lamont Colucci, American Foreign Policy Council/ Ripon College
Dr. Thomas Cooley, AFRL Space Vehicles Directorate
Timothy Cox, Tiger Innovations
Meagan Crawford, Space Fund
Dr. Karl Dahlhauser, OUSD R&E
David Davis, SMC
Casey DeRaad, NewSpace New Mexico
Marybeth Edeen, NASA JSC
Magnus Evertson, STAR Associates, Inc.
Col Eric Felt, AFRL Space Vehicles Directorate
Charles Finley, NASA

Peter Garretson, AFRL/RV
Dr. Michele Gaudreault, USSF/DST
Dr. David Hardy, Apogee support to AFRL/RV
Richard Harrison, Amer. Foreign Policy Council
Rob Hathaway, RBH Space Advisors
Gary Henry, SpaceX
Col Curtis Hernandez, National Space Council
Dale Hite, Raytheon
Lars Hoffman, Rocket Lab USA
Jeff Hurley, MDA GSI
Talbot Jaeger, Novawurks
Dr. Paul Jaffe, US Naval Research Laboratory
Johanna Spangenberg Jones, Defense Innovation Unit
Dr. Richard Joseph, USAF Chief Scientist
Susan Kaplan, Modal Technology Corporation
Matthew Keihl, NewSpace New Mexico
Chris Kemp, Astra
Jim Keravala, OffWorld
Lydia Kesatie, NewSpace New Mexico
Stephen Kitay, OUSD Policy
Lt Col Joshua Kittle, US Air Force
Katherine Koleski, Defense Innovation Unit
Maj Rachel Kolesnikov-Lindsey, Defense Innovation Unit
Bernard Kutter, United Launch Alliance
Michael Laine, LiftPort Group
Dr. Bhavya Lal, IDA Science & Tech. Policy Inst.
Juli Lawless, Made In Space, Inc.
Janna Lewis, Northrop Grumman Corp
Elaine Lim, Aerospace Corporation
Maj Timothy Locke, SAF/AQRT
Tom Loftus, Razor's Edge Ventures
Doug Loverro, NASA Headquarters

Scott Maethner, Universal Technology Company
Adrian Mangiuca, NanoRacks
John Mankins, Mankins Space Technology, Inc.
Dr. Neil McCasland, Applied Technology Assoc.
Patrick McClure, Los Alamos National Lab
CAPT Stephen Melvin, US Navy
Rob Meyerson, Delalune Space
Ric Mommer, Defense Innovation Unit
Michael Moses, Virgin Galactic
Mark Mozena, Planet
Dr. Joel Mozer, USSF Chief Scientist Office
Steve Nixon, SmallSat Alliance
Kevin O'Connell, Department of Commerce
Jose Ocasio-Christian, Caelus Partners
Kristin Paul, United States Space Force
Joseph Pauloski, Department of Treasury
Jay Penn, Aerospace Corporation
Bruce Pittman, NASA Space Portal
Dr. Dan Rasky, NASA Space Portal
Brad Reed, SMC/ZAT
Jeff Rich, Xplore Inc.
Dr. Stephen Robinson, UC Davis Center for Spaceflight Res. (*former NASA Astronaut*)
Dr. Gordon Roesler, Defense Innovation Unit
Daniel Romm, USSF SMC/ZAT
Andrew Rush, Made in Space
Dr. Mir Sadat, Defense Intelligence Agency
Tony Samp, DLA Piper
Eileen Sanchez, CA Governor's Office of Planning & Research
Mark Sarojak, GeoNeo
Dr. Grant Schaffner, Stress Engineering Services
Dr. Edl Schamiloglu, University of New Mexico
Tex Schenkkan, Defense Innovation Unit (NSIC)
Robbie Schingler, Planet
Victoria Schneider, OUSD(R&E)/R&T/RT&L
John Serafini, HawkEye 360
Dr. Joel Sercel, TransAstra Corporation
David Shogren, Treasury Department/CFIUS
Pav Singh, Defense Innovation Unit
Dr. George Sowers, Colorado School of Mines
Daniel Suarez, Author
Mike Suffredini, Axiom
Scott Suhr, Aerospace Corporation
Jeff Thoben, Quilty Analytics
Jeff Todd, USSPACECOM J8 Science, Technology & Innovation
Dr. Derek Tournear, Space Development Agency
Kathy Trimble, Council on Competitiveness
Rick Tumlinson, Space Fund
Frank Turner, Space Development Agency
Albert Varma, HQ USSF/ST
Mandy Vaughn, VOX Space
Dr. Paolo Venneri, USNC-Tech
John Wagner, Sierra Nevada Corp
Dr. William Walker, Stress Engineering Services
Micah Walter-Range, Caelus Partners
Dr. Peter Wegner, Spaceflight / BlackSky
Carol Welsch, Northrop Grumman Space Systems
Sheena Winder, AFOSR/IOA
Dennis Wingo, SkyCorp
James Winter, AFRL SSPIDR

# APPENDIX B
## PREVIOUS REPORTS & KEY RECOMMENDATIONS

**Space Power Competition in 2060: Challenges and Opportunities**
Report on the Space Futures Workshop 1A

9 Mar 2019

Distribution D:
Authorized to the Department of Defense and US DoD contractors only

- A long-term, national space strategy integrating civil, commercial and national security space lines of effort must be developed to retain the US' dominant and leadership position in the emerging future of space. This strategy must account for the possible space futures developed in the workshop.

- The overall strategy must address how the national security establishment will defend the full range of national interests in space—not just the services that are provided directly for national security.

- AFSPC must commit the resources to continue to lead in completing the remaining steps in the process to define these futures as key inputs to the strategy and to determine their implication for present and future defense strategy. US Space Command should similarly commit resources to this end as part of their strategic and operational execution missions.

- Essential capabilities and technologies to enable positive future outcomes must be developed by the whole of government. An investment, policy, and regulatory strategy must be pursued to ensure those capabilities.

- To maintain our technological advantage in space, the nation must commit to continued investment in science and technology related to the rapidly changing global space environment.

---

Download

**State of the Space Industrial Base: Threats, Challenges and Actions**
A Workshop to Address Challenges and Threats to the US Space Industrial Base and Space Dominance

30 May 2019

Distribution A:
Approved for Public Release. Distribution Unlimited.

- Upgrade of our own methodologies such as shared, trusted supply chains and interoperable technology standards that accelerate viable commercialization of the space economy.

- Develop a more flexible, US-lead markets for space capabilities that spread the risk, increase the pool of investors and establish US leadership in setting a framework for a cislunar commercial space economy that creates wealth and security with our allies and partners who share our common norms and values.

- Changes in US government procurement and licensing processes and other regulations to eliminate unnecessary delays and micro-management of the space industrial base's ability to rapidly deliver next generation space capabilities and to enable early US investment in capabilities from emerging, innovative elements of the space industrial base

### The Future of Space 2060 & Implications for US Strategy
Report on the Space Futures Workshop

5 Sep 2019

Distribution A:
Approved for Public Release. Distribution Unlimited.

[Download]

- The 2060 space world will be highly complex and diverse as to the number of state and non-state actors, their capabilities, and their interests.

- Commercial space presents unique issues as to ownership and sovereignty that, if not resolved, could lead to commercial space entities as independent or semi-independent space powers, resulting in significant opportunities and challenges to US space power.

- Space power will be widely distributed, making it impossible for any one nation or entity to have predominant space power in the civil, commercial, and military domains.

- The diversity and distribution of space power enables a wide range of alliances, partnerships, and shared interest. These relationships will be diverse and vary with time as the interest and capabilities of space faring entities develop and change. This complexity poses significant challenges to the US to maintain and exercise the space power needed to protect its interests in space and in the terrestrial sphere.

# APPENDIX C
## METHODOLOGY

This workshop builts on the analytical approach and results of the previous USSF and AFRL/DIU co-sponsored workshops, and the present expressions of our Nation's space policy. The workshop's goal was to perform the next level of decomposition of the state, issues and challenges to the US space industrial base so as to:

1. Identify the key inflection points that affect the trajectory of the future of global and US commercial space capabilities.
2. Determine the specific actions by the defense community and across the US whole-of-government with the commercial space community to ensure a space future that guarantees the space capabilities and power to protect US national interests.
3. Determine the impact of the coronavirus pandemic on the US space industrial base and short-term actions required to mitigate that impact.

The workshop examined the areas of expected commercial growth along the 6 lines of commercial activity and development in the approximate order of how they may grow in economic importance over time, as US activity expands across cislunar space and beyond,

1. **Space Information Services** – Space communications/internet, PNT and the full range of earth observing functions which have commercial, civil and military applications. This includes the maximum commercial exploitation of the electromagnetic spectrum to, in and from space for information services.
2. **Space Transportation and Logistics** – Transport to, in and from space, and the ability to service, repair, upgrade and construct systems in space.
3. **Human Presence in Space** – Human presence as part of exploration, space tourism/personal space travel and to support the full range of commercial space enterprises.
4. **Space Power** - Power in, across and from space.
5. **Space Manufacturing and Resource extractions** – Space manufacturing and resource extraction for terrestrial markets and to provide space infrastructure and buildout of greater commercial space.
6. **Space Policy and Finance** – Government space policy and financial incentives that affect commercial space enterprises.

First, the findings and challenges from the three previous workshops were summarized, including a review of the competitive environment the US will face in the future across the full range of future space rivals and adversaries. This was followed by a panel overview of each of the six areas. Breakout groups subsequently explored each topic in-depth for the remainder of the workshop , interspersed with plenary sessions to promote cross-fertilization of ideas among the subgroups. Each breakout-group addressed the following six questions.

1. What is the current state of US commercial capabilities and investments in the area?
2. What are the specific connections and implications for this area to the needs and responsibilities of national defense?

3. What are the present and future challenges to US commercial capabilities and investment in that area both technologically and competitively with other present and emerging space powers?
4. What are the key inflection points in terms of events or breakthroughs that will most affect US commercial capabilities and investment in this area?
5. What defense and overall government options are there to affect those events or accelerate those breakthroughs to the advantage of the US?
6. What decisions or actions can the national defense community and all-of-government in partnership with industry take to affect these events or accelerate those breakthroughs to the advantage of the US? These were specifically addressed in the short-, mid-, and long-term, actions within DoD and more broadly across all of government. Specific attention was given to short-term actions that can be taken to mitigate the impact of the coronavirus epidemic.

# APPENDIX D
## COVID-19 IMPACTS TO THE SPACE INDUSTRIAL BASE[63]

## EXECUTIVE SUMMARY

Thirty-five US commercial space companies were surveyed to determine impacts of the COVID-19 pandemic and economic downturn that followed:

- Median number of employees: 77
- 60% are venture backed
- 38% are eligible for Small Business Administration (SBA) assistance although Paycheck Protection Program (PPP) is of limited help due to highly specialized, highly compensated workforce
- 34% were significantly impacted by the OneWeb Chapter 11 bankruptcy filing
- All currently have one or more contracts with the Department of Defense (DoD)
- Total number of suppliers reported: 4,525 (98% are US-based)

Most companies are experiencing elevated risk requiring decisive action:

- 30% immediately at risk
  - Fast growth stage relies more on innovation & investment than revenue
  - These companies have high burn rates and insufficient cash on hand to ride out the crisis
  - Most companies were actively raising capital when the economic down-turn occurred
  - Commercial sales pipeline frozen as companies assesses impact and slash orders
- 54% at moderate risk and hunkering down
  - These companies have cash on hand but are scaling back operations to survive
  - This group will likely weather the current crisis but remain highly dependent on critical supply chains that are only one or two vendors deep
- 16% at low or uncertain risk due to reduced demand for production
  - These companies are highly dependent on the at-risk manufacturers identified above
  - Most are diversified or qualify for SBA assistance in the short-term
  - An extended slowdown in the sector will put critical suppliers in distress

Recommendations:

- Stimulate private investment in the US space industrial base through DoD's immediate actions of awarding new contracts or modifications to scale prototypes & procurements
  - Prioritize hardware-focused launch vehicles and spacecraft to stimulate critical US-based supply chains
- Temporarily relax SBIR/STTR regulatory restrictions to enable contracting flexibility
- Fund National Security Investment Capital (NSIC) to incentivize private investment in dual-use hardware and shunt foreign influence in early stage startups of interest to national security and defense

---

[63] Survey responses reflect the views of the participating companies representing the US space industrial base and do not necessarily reflect the official position of the USSF or Department of Defense.

- Accelerate actions required to form a US-based Space Commodities Exchange to diversify and grow investments, increase the number of successful exits, and increase resiliency during future disruptions to the US economy

## INTRODUCTION

The impact of the novel Coronavirus Disease 2019 (COVID-19) pandemic on the United States has been profound. Within days of a national emergency declaration, the US economy stalled as widespread shelter-in-place orders were implemented resulting in rising unemployment and unprecedented disruptions to normal economic life. The capital markets that fuel commercial innovation have contracted leaving many innovative technology companies caught off guard with little to no cash to sustain operations. Nowhere has this been more evident than in the burgeoning US space economy. Soon after the first SpaceX commercial launch successfully delivered a satellite to low Earth orbit in 2009, the new space economy has flourished with more than $25.7 billion of private investment globally. By the end of 2019, the US space workforce had grown to 184,000 in core-employment - a 4.1% increase over the prior year, with a record $5.8 billion of capital directed to new space products and services.[64,65] More than 70% of this 2019 investment was sourced from venture capital.[66] The US space economy is still nascent and highly dependent on early adopter customers which adds to its vulnerability.[67] However, its size, scope and impact on the national security innovation base warrant special attention[68]. From a national security perspective, space is not just contested militarily but economically as well. Notably, 34% of the fourth quarter 2019 global investment in space went to relatively new launch companies based in China.[69] The health of the US space industrial base for the foreseeable future will be highly dependent on actions our nation chooses to take (or not) within the next 90 days. As Defense represents a significant portion of the space marketplace, focused procurements of hardware-centric products and services are recommended to prevent the sector from falling into a prolonged state of hibernation and starving off a highly specialized, US-based supply chain - a critical resource and talent base that would likely require decades to reconstitute.

This appendix summarizes a survey of commercial space companies performed between 1 and 17 April 2020. These companies represent a small but well-funded fraction of the US space industrial base from small launches to satellite services, ground systems to advanced analytics platforms, established corporations to new start-ups. All currently hold one or more competitively awarded contracts for commercial products, services or prototypes supporting national security through the Department of Defense (DoD). In keeping with the US National Space Strategy, the dynamic and cooperative

---

[64] US Bureau of Statistics (2019).
[65] Space Foundation (2020). The Space Report 2020, Q1.
[66] Messier, D. (2020). Soaring Investment in Commercial Space Dominated by a Handful of Companies. Parabolic Arc.
[67] A nascent market is defined as a very new and formative market; the market development stage in which vendors sell their products or services to innovator and early adopter customers. In typical nascent markets, rapid innovation occurs, many new competitors (including startups) enter, and competition tends to revolve around innovation and product features rather than around brand, service or price.
[68] The national security innovation base comprises the ecosystem of capital, research, knowledge, capabilities, policies, incentives, and people that turns ideas into innovations and transforms discoveries into useful technology and products to protect our national security.
[69] Space Angels (2020). Space Investment Quarterly Q4 2019.

interplay between the national security, commercial, and civil space sectors is absolutely paramount during this crisis.

## STRATEGIC IMPERATIVE

In 2018, US sources of commercial innovation were recognized in the NDS as critical to sustaining the Department's technological advantage. Establishing an unmatched twenty-first century national security innovation base that effectively supports Department operations and sustains security and solvency is a key objective of the NDS.[70]

> *"Small businesses may feel the economic pain the most, and that could spell trouble down the road. A 2018 Pentagon report on the defense-industrial base warned of "domestic extinction" among the sole suppliers for critical industrial parts — shops that could fold under rough economic conditions as the world is seeing now."[71]*
>
> *Aaron Mehta, Defense News*

Economic leadership is foundational to sustaining military superiority in the space domain. US adversaries understand this and have embarked on a path to disrupt US preeminence with both military and economic instruments of national power. The economic downturn that soon followed the COVID-19 pandemic has severely impacted the availability of investor capital sustaining the US space economy. The DoD has leveraged this capital investment in space for years, and at a ratio exceeding 30:1.[72] In other words, for every dollar of DoD funding applied to the prototype or procurement of a commercial space product or service, thirty dollars of development was funded by non-government sources—primarily venture capital. The US government benefits from this leverage in two ways: (1) more research and development funding is available for modernization priorities that are not inherently commercial; and (2) the speed at which commercial products and services evolve and improve are not constrained by the discontinuous and unpredictable budget cycles of the federal government. The combination of these benefits have fueled US sustained leadership in commercial space. Should no action be taken, and the US new space economy be allowed to falter, future capital investment will be much more difficult to attain and an even greater dependence on government funding will be required to reconstitute capabilities lost.

Timely and prescriptive solutions are required to preserve the US commercial space industry investment; $14 billion would be required to reproduce the existing investment.[73] However, we are not suggesting a wholesale bailout of every vendor in the supply chain. A blanket bailout absent deliverables is an inappropriate remedy for this problem. The best remedy is one that energizes new orders for parts and materials, relies on suppliers and subcontractors that stretch across the United States, and delivers new commercial capabilities to the space domain that enhance and complement US

---

[70] United States. *Summary of the 2018 National Defense Strategy of the United States of America.* Washington D.C: US Department of Defense.
https://dod.defense.gov/Portals/1/Documents/pubs/2018-National-Defense-Strategy-Summary.pdf
[71] Mehta, A. (2020). How coronavirus could impact the defense supply chain. DefenseNews.
[72] Based upon DIU's analysis of space portfolio investments.
[73] The US share (55%) of the total investment in new space since 2009 ($25.7B).

national security and defense. An ideal solution is one that: (1) remains aligned with established national security priorities; (2) sustains a demand signal for commercial goods and services that reassures the capital markets and the downstream supply chain; and (3) provides sustained revenue through the recovery period versus a short-term fix that encourages staffing and production cuts, all but assuring the stifling of innovation and starvation of the supply chain.

Speed is paramount, but purposeful and focused investment is essential. Contract awards and modifications that accelerate and scale US commercial solutions involving small launch, small satellites, reusable spacecraft, ground systems, information services and their respective supply chains will serve as a profound countermeasure to the effects resulting from COVID-19. The interdependence of economic and military power mandates steadfast action to preserve the US space industrial base with the same vigor taken to establish the USSF. Focusing on one without the other is insufficient.

## IMMEDIATE FOCUS

Objectives for financial assistance of the US space industrial base:

1. Keep the US commercial space workforce working (and healthy)
2. Keep supply chains active, viable and US-based
3. Minimize delays & cancellations of new products or services of interest to the DoD
4. Expedite transactions using other transactions and other accelerated agreements
5. Mitigate predatory foreign investments, acquisitions and intellectual property transfers
6. Keep the US leading and winning in space for long-term economic and military advantage

## SURVEY

This survey was initiated on April 1st approximately two weeks after the President declared a national state of emergency concerning the COVID-19 outbreak.[74] Participating companies are fairly well funded with familiar names and <100 employees (median value). The vast majority provided detailed information on their suppliers. The survey questions were formulated to assess risk, determine company intent and scope the ramifications on the greater US space supply chain (see Annex B-1). Ninety percent of the surveyed companies responded with detailed information.

Upon analyzing these responses, it was determined that a follow-up survey was required to ensure its observations were correct and recommendations complete (See Annex B-2). The follow-up survey was initiated on April 10th, a little more than a month into the national emergency. Again, more than 90% of the companies responded. This was the first time that companies were specifically asked about potential impacts resulting from the OneWeb Chapter 11 bankruptcy filing on March 27th.[75] The follow-up survey also asked about Small Business Administration (SBA) benefits eligibility, applications and receipts under the Coronavirus Aid, Relief, and Economic Security (CARES) Act; and more detailed requests on the status of funding, investor relations, layoffs, furloughs and future planned reductions in employees; and critical impacts within their supply chain.

---

[74] White House (2020). Proclamation on Declaring a National Emergency Concerning the Novel Coronavirus Disease (COVID-19) Outbreak.
[75] Henry, C. (2020). OneWeb files for Chapter 11 bankruptcy. SpaceNews.

## FINDINGS

As illustrated in Figure B-1, most of the US commercial companies surveyed are experiencing an elevated risk that generally fall into one of three categories:

**CAT 1: Immediately at Risk** (Cash flow required by 30 Jun 2020) - These companies were caught in the middle of, or just prior to, a capital funding raise. Capital markets have severely tightened, and now these companies are struggling to sustain their existence. Most are hardware-oriented firms with large numbers of employees and extensive supply chains. In fact, more than 90% of the capital investments in space over the past decade have been focused on hardware. This includes several rocket companies, satellite and spacecraft manufacturers. This is also where the bulk of the highly specialized employees go to work in the commercial space industry. Companies with the highest growth potential typically take smaller, incremental raises and step up valuations between raises by meeting aggressive milestones. This group has the shortest life expectancy due to a deficiency of cash and is the most at risk for predatory foreign investment or acquisition.

**Figure D-1:** Elevated Risk Profile for the US Commercial Space Industrial Base post COVID-19

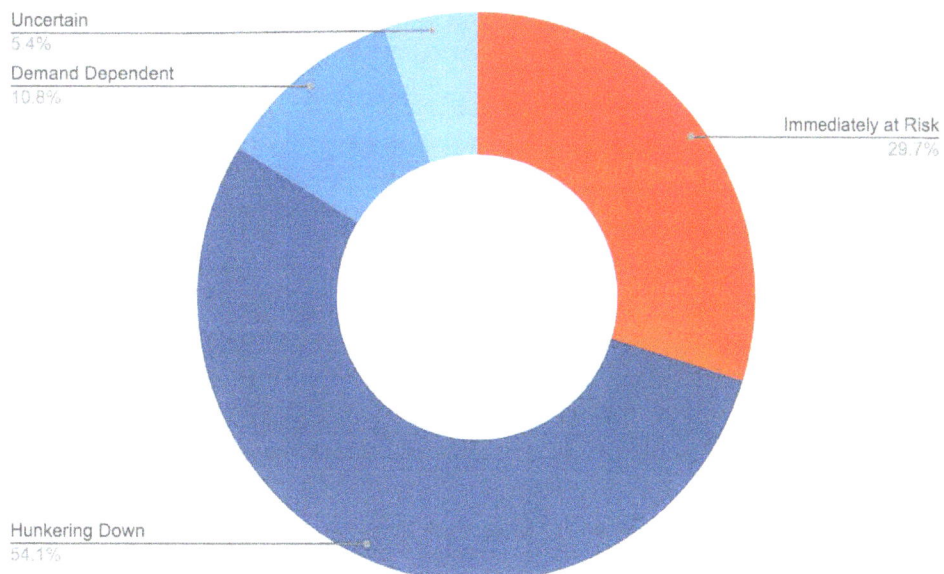

**CAT 2: Hunkering Down to Extend Runway** (Cash flow required by 31 Dec 2020) - These companies have some cash-on-hand due to a recent funding round or banknote but have expressed uncertainty in their opportunity pipeline. In response, these companies are beginning to slow down or cease new product offerings to preserve resources for the duration of the economic downturn. Many satellite companies fit in this category. At least one was planning to place $10 million in new orders for long lead items from suppliers when the COVID-19 national emergency was declared and, subsequently, those plans were shelved. Most are suspending satellite production and canceling commercial launch reservations which has a domino effect on the highly specialized suppliers who depend on them for business. As uncertainty extends later into the year, many of these firms have indicated that they will likely reduce the workforce in order to sustain operations.

**Table D-1:** Summary of COVID-19 Impacts Survey

**A Sampling of Commercial Companies representing the US Space Industrial Base**

1-May-20

*This product is updated daily to reflect new information received from survey participants regarding staffing, supply chain, finance or CARES Act relief*

| Company Info | Personnel Impacts to Date | Supply Chain | Capital Financing | CARES Act | |
|---|---|---|---|---|---|
| # Surveyed | # Furloughs | # Suppliers | % VC/PE Backed | % SBA Eligible Companies | |
| 37 | 102 | 4,534 | 57% | 38% | |
| % Responded | # Layoffs | % Companies Reporting Suppliers | Ratio of Private Investment | # SBA Applications / Payments | |
| 89% | 78 | 76% | 60:1 | 10 | 4 |
| Median # of Employees | # Future Reductions Planned | % Suppliers US-based | NSIC Eligibility (If Funded) | SBA Relief Funds Rec'd to Date ($M) | |
| 76 | 93 | 98% | 27% | $0.5 | |

*Factors considered in assessing risk: cash-on-hand, Burn-rate, cancelled orders or services (i.e. launch), Impacts resulting from OneWeb bankruptcy*

| % Immediate Risk | % Medium Risk | % Low Risk | % Uncertain Risk | % Impacted by OneWeb Ch 11 |
|---|---|---|---|---|
| 30% | 54% | 11% | 5% | 32% |

| What are your projected cash flow requirements through 2021? | | | Anticipated investor response to stimulus 'signal'. Cost to rebuild US Sector? | |
|---|---|---|---|---|
| Cash flow req'd by 30 June ($M) | Cash flow req'd by 31 Dec ($M) | Total req'd through 2021 ($M) | Est. Leverage Value ($M) | Est. Replacement Value ($M) |
| $291 | $674 | $965 | $3,859 | $14,135 |

*(1) National Security Innovation Capital (NSIC) was approved in the FY18 NDAA to facilitate early stage capital investment in dual-use hardware startups. (2) The Ratio of Private Investment is a measure of private funding applied to development costs compared to the value of the DoD prototype contract. (3) Leverage Value assumes US investors respond strongly to a signal sent by a focused DoD procurement of commercial products & services. (4) Replacement Value represents 55% (US share) of the total private equity and venture capital investment in commercial space since 2009. (Source: Space Angels, 2020)*

**CAT 3: At Risk Due to Reduced Development and Future Production Demand in the Sector** (Cash flow safe at present but uncertain beyond 2021) - This group includes bootstrapped companies that share a dependency with those represented in CAT 1 & 2 above. These companies tend to be smaller suppliers with cash-on-hand from existing government business. Most are eligible for short-term relief from the SBA under the Coronavirus Aid, Relief, and Economic Security (CARES) Act. Unfortunately, long term prospects for some in this group remain uncertain as future success is tied to a growth in demand, or more specifically to the orders received. Many of these companies are highly specialized suppliers in a pool of US manufacturers that is extremely shallow, often one company deep in particular specialties. They compete with offshore alternatives, many of whom are not as critically affected due to large government subsidies or State sponsorship.

## OBSERVATIONS

**In a software-centric innovation economy, commercial space is hardware-centric with 98% of the supply chain based domestically despite strong competition offshore.** The map in Figure B-2 below illustrates the density and distribution of US space companies, their subcontractors and suppliers across the United States. Of the companies surveyed, 74% reported their suppliers by name or locality. Supply chains provide a proprietary advantage within the competitive commercial industry, so it is remarkable that this level of detail was obtained. California is home to more than 1,600 of these suppliers and the concentration in California has increased by three times the number that existed in 2014 according to the US Department of Commerce.[76] A more comprehensive survey to fully scope

---

[76] Botwin, B. (2014). US Space Industry 'Deep Dive' Assessment: Small Businesses in the Space Industrial Base.

the extent of the US space industrial base is underway by the US Commerce's Bureau of Economic Analysis (BEA).[77]

**Figure D-2:** Density map of US space companies, their subcontractors and their key suppliers.

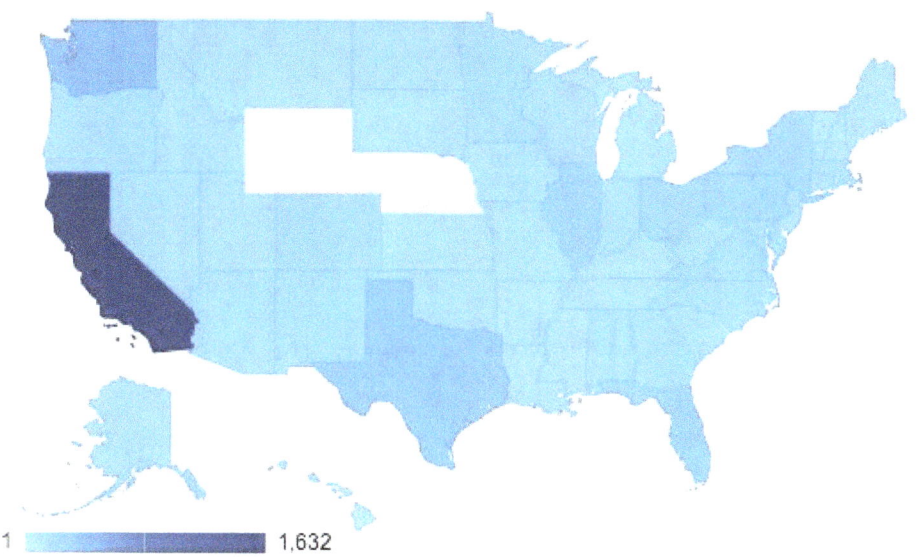

**OneWeb's bankruptcy will have damaging ramifications for the commercial space industry's already-fragile supply chain.**[78] OneWeb's Chapter 11 filing has placed key suppliers under duress which is now further amplified by the events surrounding the COVID-19 crisis. One third of the companies surveyed were impacted directly or indirectly by the OneWeb bankruptcy. Most notably, key suppliers to many of these companies had scaled operations through capital investments necessary to feed the OneWeb satellite production line. Others had anticipated providing launches and other services with revenues to commence in 2021. The untimely demise of OneWeb could not have occurred at a worse time. When combined with the COVID-19 crisis, these factors will result in cascading impacts across the US space industrial base (see Figure D-3).

---

[77] Highfill, T. et al (2019). Measuring the Value of the US Space Economy. BEA.
[78] Quilty, C. (2020). OneWeb hits BK - What's Next. QuiltyAnalytics.

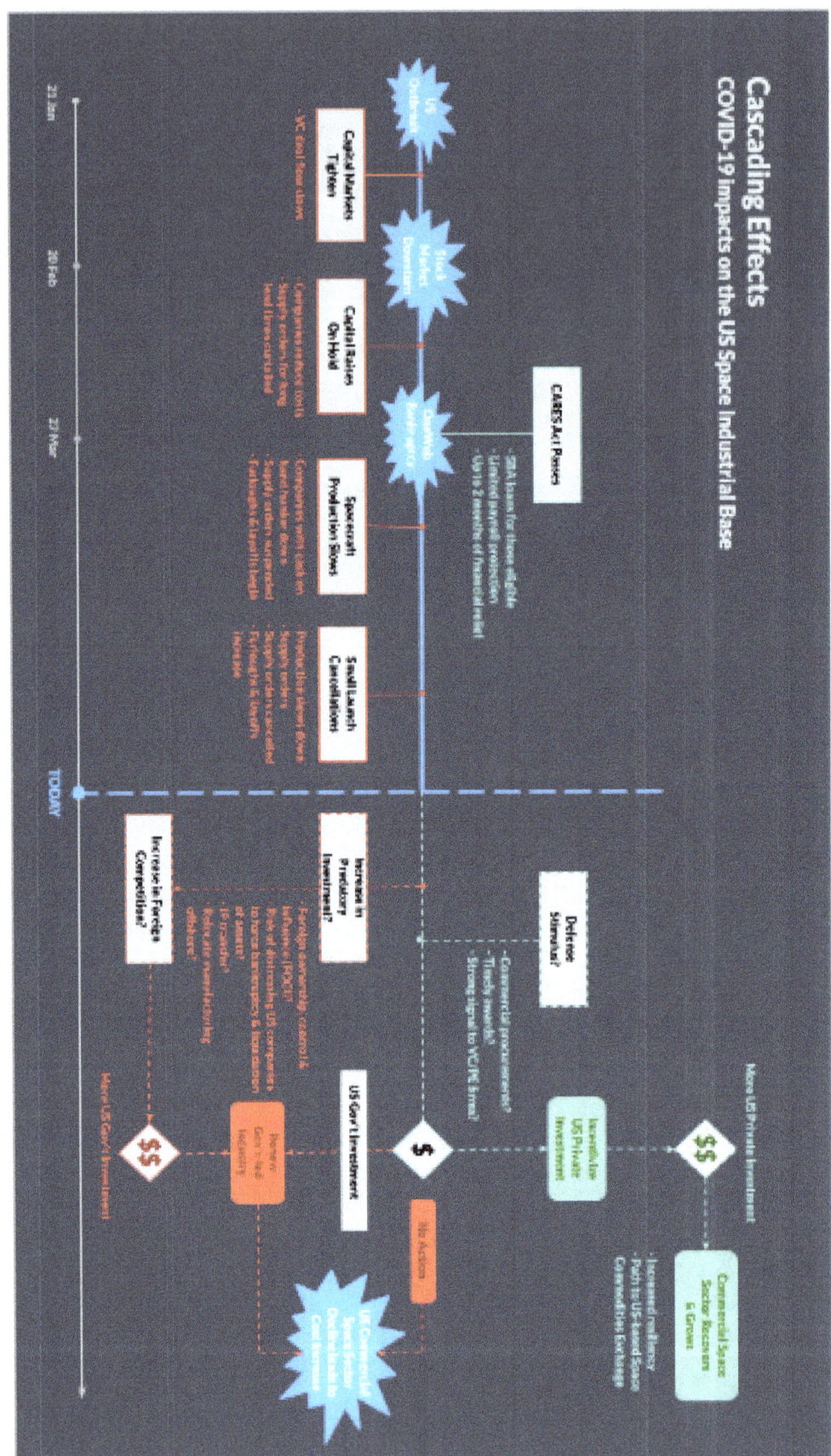

**Figure D-3:** Cascading effects resulting from the COVID-19 pandemic and economic downturn.

**The US may forfeit its leadership in commercial space.** The new space economy is a product of US ingenuity and entrepreneurship. Since 2009, the US has led the commercialization of geospatial imaging via small satellites using advanced analytics to derive information from satellite data autonomously, and reusable rockets that have significantly reduced the cost of placing mass in orbit. More than 500 companies fueled by nearly $26 billion define today's new global space economy. At the outset of the COVID-19 crisis, the US controlled 55% of this market with $14 billion of private investment over the past decade. Should the commercial space sector collapse domestically, it will likely be supplanted by foreign entities who benefit from more direct government subsidies and foreign national strategies such as China's civil-military fusion. The cost required to stabilize the US commercial space industry now may be significantly less than the cost of replacing it.

**US investors will respond favorably to a 'strong signal' conveyed by focused DoD procurements of commercial space products and services.** Despite the economic downturn, there is still capital available and new investment deals underway. According to PitchBook, sixty-six venture capital deals were recorded in China for the week ending 28 March 2020, the most of any week this year and nearly the same number recorded at this time last year.[79] PitchBook also reported that venture capitalists held $188.7 billion globally equating to 2.5 years of capital on hand as of mid-2019. Investors are looking to reduce risk and increase protection of their assets during the post-COVID-19 recovery. DIU has spent five years developing trusted relationships with venture capital and private equity investors located in Silicon Valley and elsewhere. Over much of this time, one sentiment has remained steadfast: it is better to issue fewer but more meaningful contracts in order to elicit a response from quality US investors.

---

*"Our industrial base is going to be massively stressed. We are going to have to work in a very creative and aggressive fashion to try to protect our industrial base in a broader way, without question."*

*Adam Smith, HASC Chairman*

---

**Thirty percent of the companies surveyed would benefit from National Security Innovation Capital (NSIC) in the development of dual-use hardware or manufacturing solutions.** Today, less than one third of all investments by US venture capital firms (and less than 10% of investments in information technology) are in hardware due to the higher levels of technical risk and longer development cycles relative to software. This is a particularly troubling revelation as the nation grapples with the realization that much of its critical manufacturing base has been moved offshore to China. NSIC is an authorized but currently unfunded program to incentivize private investment in dual-use hardware startups.[80] It does so by creating a public-private partnership program that delivers small but catalytic government funding to attract equal or greater venture capital investment by reducing risk and providing a strong market signal of importance and opportunity. Accelerating the initial funding to NSIC would provide a beneficial second lever to scale venture funding to at-risk companies.

---

[79] Thorne, J. (2020). China's VC industry bounces back after coronavirus-induced winter. Pitchbook.
[80] Authorized in §230 of the 2019 National Defense Authorization Act (NDAA) and based in part on §1711 of 2018 NDAA.

**China is the most likely benefactor of a collapsed US space industrial base**. It is ironic to consider that the nation most responsible for the conditions leading up to the novel coronavirus global pandemic stands to benefit the most from its impact on the US space industrial base. China has relentlessly pursued US sources of commercial innovation both legally and illegally.[81] A 2019 report co-authored with the Air Force Research Laboratory identified the challenge and threat of China to the US space industrial base to be significant and growing.[82] Once technical parity is achieved, China combines government resources and predatory pricing to achieve a dominant global market share position. This has proven true in other technology markets including small drones, photovoltaic cells and potentially 5G communications.

**Figure D-4:** China has already surpassed the US in number/share of space launches per year.[83]

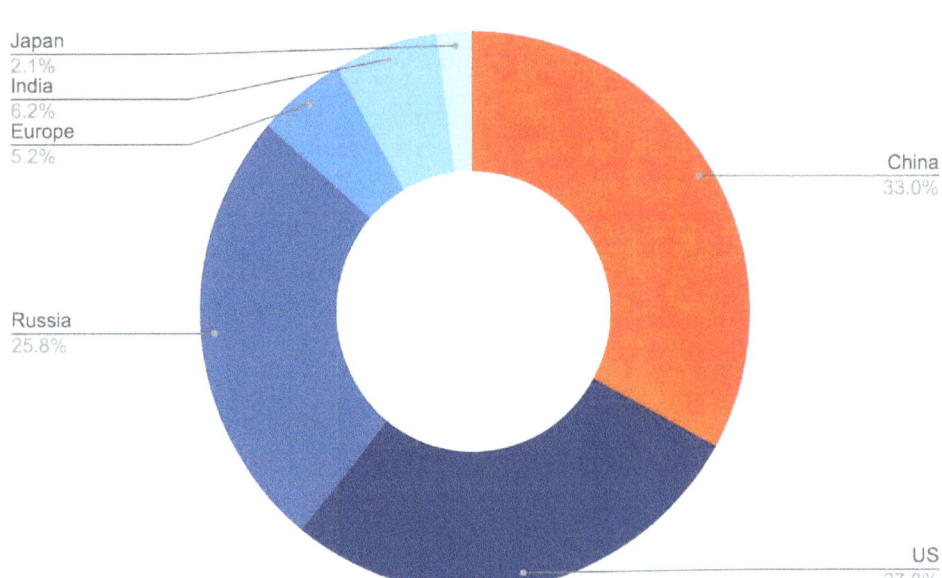

The rate and scale of investment in Chinese space companies poses a clear and present danger to the sustained US leadership in the commercial space sector.[84] There will likely be no Sputnik moment involving China's rise as a space power. They are on a patient, methodical trajectory to surpass the US as the dominant space power as illustrated in Figure B-4 above.

## THE CASE FOR STIMULATING THE US SPACE INDUSTRIAL BASE

The following factors make investment now in the US space industrial base imperative.

**The US space workforce is highly specialized and not easily reconstituted**. Engineers and scientists typically earn >$100K salary per year. Once layoffs or shutdowns occur, it will take these

---

[81] Rohrlich, J. and Fernholz, T. (2019). China is trying to steal military space tech. The US is running stings to stop it. Quartz.
[82] Cooley, T. et al (2019). State of the Space Industrial Base: Threats, Challenges and Actions. DoD.
[83] Parsonson, A. (2020). Rocket Rundown: A Review of Spaceflight in 2019. Rocket Rundown.
[84] Beall, A. (2019). China's private space industry is rapidly gaining ground on SpaceX. Wired.

companies considerable time to rebuild the talent pool required to restore full operational capabilities, if that is even possible.

**More than half of the private investment in commercial space is sourced from venture capital.** This has been the case since 2012 (see Figure B-5). Preserving access to this funding stream is paramount to sustaining US competitive advantage in space.

**Figure D-5:** Diversity of private investment in commercial space. (Source: Bryce Space & Technology)

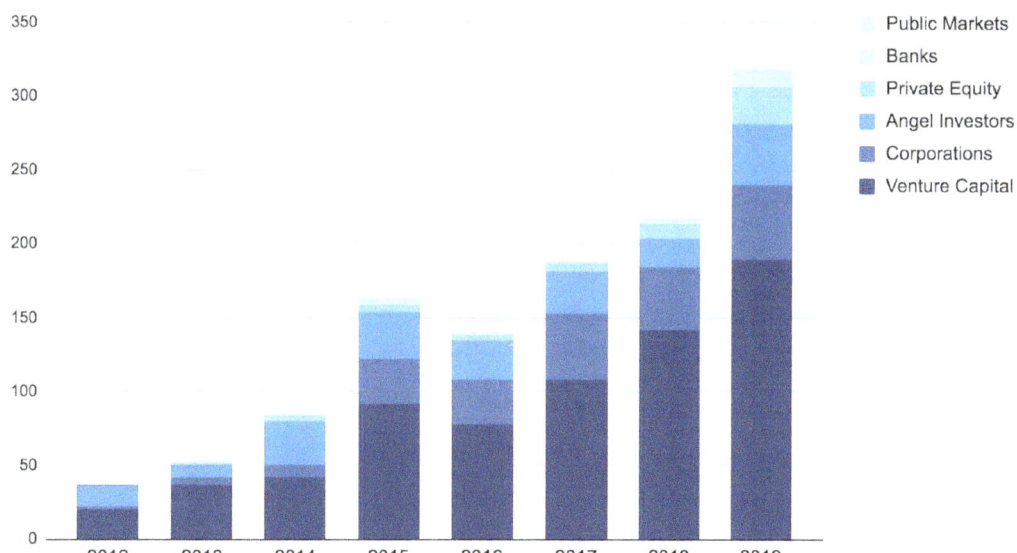

**Commercial space technology preserves and sustains both civil and military leadership in the space domain.** Commercial products, services and infrastructure may be mobilized for national security and defense when needed in the future much as was done with the commercial aviation industry at the onset of WWII.[85]

**The US space economy is nascent but growing toward commoditization.** The demand for commercial space services not solely dependent on the government is beginning to emerge. With it, the path to commoditization of commercial imagery sources, and the processed information derived from them, will lay the foundation for market growth and more diversified financial investments. US sustained leadership will ensure that a future space commodities exchange is US-based with its growth strengthening the value of the US dollar.

## COURSES OF ACTION

The elevated risk confronting commercial space companies (as reflected in Figure 1) warrants careful consideration of potential courses of action (COA) by the US government, and more specifically the DoD, in response to the COVID-19 crisis:

---

[85] Today's space industrial base is reminiscent of the early commercial aviation industry during the interwar period (1920 to 1940). The US government was a strong early adopter customer during these formative years which enabled an accelerated pivot to attain air power during WWII and the rise of the US as a great economic power afterwards.

## COA 1: Decisive Action to Support Commercial Suppliers and Encourage Private Investment

Focused and prescriptive government action in the form of accelerated contract awards or modifications for prototype projects, products and services that address priorities or inform future procurements by the USSF, the Services and other defense agencies with equities in the space domain. This COA embraces innovation processes already in place by prescriptively increasing funding and deliverables on existing contracts with commercial companies already vetted through established acquisition processes such as DIU's Commercial Solutions Opening. This COA will ensure scalable delivery of launch vehicles, spacecraft, ground systems and other items intended to stimulate much needed activity in the US space supply chain.

> *"The single most powerful thing DoD can do to preserve the innovation supplier base is to continue to purchase goods and services during the COVID-19 economic downturn. For a startup, purchases show that the startup's technology is valuable and that customers want it. Investors will continue to fund startups whose technology sells. These investors will drop startups that do not have sales, and in a poor economy they will drop them even faster. As a leading customer (arguably the leading customer) in the space industry, the DoD has the power to show unequivocally that a startup's products and services are valuable."*
>
> *Nontraditional Defense Innovator*

## COA 2: No Action

Allow the free-market forces to determine the fate of the nascent commercial space sector and accept that a winnowing of the US companies will be accelerated by events surrounding COVID-19. This COA recognizes that some companies and their technologies will fall to foreign investors and adversaries; others with cash on hand will weather the crisis albeit in a weakened condition and, as a result, US innovation will be temporarily stalled allowing foreign competitors much-needed time to leap-frog American capabilities.

## COA 3: Renew Exclusive Government Funding

Propose renewed and strengthened focus on government financing to serve the needs of national security and defense. This COA addresses the commercial viability risk incurred during an economic downturn but at significantly greater cost to the taxpayer over time. The pace of government-led innovation is influenced by political priorities and discontinuities in defense budget cycles. As an example, there have been on average 4.6 continuing resolutions in funding the US government per fiscal year since 1977.[86] In 2018, the GAO found that continuing resolutions have a significant adverse impact on contracts and grants.[87] Incentives would be necessary to attract and retain talent coupled with significant deregulation to ensure competitiveness. Defense already constitutes roughly half of all US government discretionary spending each fiscal year. In contrast, the US economy has sustained a net growth over the past three decades fueled by commercial innovation despite intermittent crises such as the dot-com crash in 2000 or the financial crisis of 2008.[88] We are now seeing the benefit of increased

---

[86] Wezerek, G. (2018). 20 Years Of Congress's Budget Procrastination, In One Chart. FiveThirtyEight.com
[87] Krause, H. (2018). Continuing Resolutions and Other Budget Uncertainties Present Management Challenges. GAO.
[88] Li, Y. (2019). This is now the longest US economic expansion in history. CNBC.

competition at lower-cost in the commercial space industry as illustrated in Figure B-6. In 1961 the Chairman of the Board for General Electric, Ralph Cordiner warned against the bureaucratic tendency to foster government-owned industry. To do so is emblematic of the State-controlled, regimented societies that the United States' competes with for leadership in space.

**Figure D-6:** Comparison of R&D spending by US business versus the Federal Government.[89]

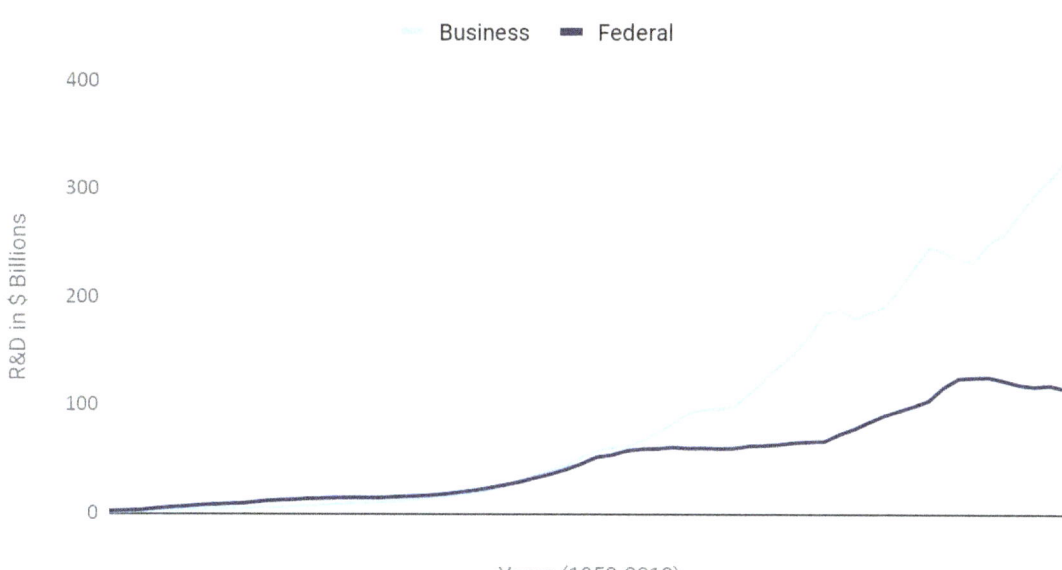

## RECOMMENDATION

**COA 1, Decisive Action to Support Commercial Suppliers and Encourage Private Investment**

Put the US space industrial base back to work now by executing timely contract awards and modifications that scale prototyping, procurement and production efforts. The space economy attracts investment capital (broadly debt and equity) by broadcasting its demand curve for the products and services that companies produce.[90] The US government can signal to private investors that it is a reliable customer for what we already committed to achieve in space.

## ADDITIONAL RECOMMENDATIONS

**Temporarily relax SBIR/STTR regulatory restrictions** such that a portion of available funds may be used in combination with RDT&E employed on Other Transactions agreement modifications and awards through FY21. Encourage broad adoption of DIU's Commercial Solutions Opening to accelerate commercial prototype awards with fair and open competition[91].

---

[89] National Science Foundation, National Center for Science and Engineering Statistics (NSF/NCSES), National Patterns of R&D Resources (annual series), February 27, 2019.
[90] Cahan, B. (2020). Stanford GSB / Urban Logic.
[91] Two commercial space prototype agreements awarded by DIU were accomplished within 57 and 60 days of the solicitation closing date and were finalized remotely during the COVID-19 Shelter-in-Place restrictions.

**Apply supplemental defense funding to preserve and protect the US space industrial base** as a critical component of both the civil space program and the national security innovation base. This funding should explicitly be used to accelerate, not slow down or stall, the advancement of commercial capabilities that retain and strengthen US leadership in space.

**Request appropriation for NSIC** be included in future omnibus and FY21 budget. NSIC can be an effective tool to seed early stage technology of interest to DoD, and to displace or thwart predatory foreign investments and acquisitions. OUSD R&E previously requested $75 million in the FY20 budget but it failed to receive an appropriation from Congress.

**Fund the preparatory actions required to form a US-based Space Commodities Exchange** to diversify and grow investments, increase the number of successful exits, and build resiliency against future disruptions to the US economy. This recommendation was adopted by the US China Commission and included in their 2019 Report to Congress.[92] An interagency effort led by US Commerce with the participation of DoD, NASA, NOAA, USGS, and other government agencies is exploring the path to commoditizing space products and services.

---

[92] https://www.uscc.gov/annual-report/2019-annual-report

# ANNEX D-1

## SURVEY QUESTIONS - APRIL 1ST

1. What are your company's immediate cash flow needs beyond the scope of current DoD contracts?
   a. Is your company eligible for relief from the Small Business Administration under the Coronavirus Aid, Relief, and Economic Security (CARES) Act? If not, why?
   b. Are current SBA offerings under the CARES Act sufficient to address your needs?
   c. With respect to existing work with DoD, have you engaged with your DoD contracts officer for any other DoD entity, to discuss the potential need for modifications or accelerated milestone payments?
2. What are your projected long-term funding needs through December 2021?
   a. If new or incremental funding were available, how would you envision it being executed in alignment with your current prototype or procurement effort?
3. Total number of employees (full and part-time)
   a. Breakdown by location, if applicable
   b. Have you laid-off or furloughed any employees since 13 March 2020 when the COVID-19 national emergency was declared?
4. Total number of suppliers
   a. Breakdown by location
   b. Are they exercising the Essential Critical Infrastructure Workforce guidance we shared with your company?
5. Optional information which may help us justify the immediacy of potential action:
   a. Last funding round (date, amount)
   b. Cash on hand
   c. Monthly burn rate (payroll, G&A, capital expenditures)
   d. Supplier base breakdown, as mentioned above

# ANNEX D-2

## FOLLOW-ON SURVEY QUESTIONS - APRIL 10TH

6. With respect to Impact: How has the recent OneWeb Chapter 11 bankruptcy filing impacted your company or a key supplier?
    a. Are there any other suppliers that you would consider critical to your business that appear to be in distress or have dramatically reduced operations/commitments?
7. With respect to Short-term relief: Has your company either (1) initiated an investigation into, (2) applied for, (3) received, or (4) determined that you would not qualify for an SBA-authorized bank loan or payroll protection relief authorized by the Coronavirus Aid, Relief, and Economic Security (CARES) Act?
8. With respect to Supplier Impact: If you have not already provided details on your supply chain, it would be extremely helpful to do so now. In order of priority, we would respectfully ask for (1) supplier zip code (preferred) or state, (2) supplier name (if you feel comfortable supplying this) and (3) an indication if you would consider a given supplier "critical" to your business. Note: If you do not wish to provide supplier names, we understand - this is proprietary information and may be considered strategic to your business; please replace it with "not disclosed" if this is the case. To be completely transparent, our intention is to identify suppliers that multiple companies deem "critical" to help alleviate a ripple effect should they go into distress.
9. With respect to Human impact: Has your company already implemented or made plans to implement near-term reductions in personnel as a result of the COVID-19 pandemic? If so, please specify total furloughs or layoffs as appropriate [Note: our current numbers appear to be under-reported which means we may have to omit the human dimension of COVID-19 impacts].
10. With respect to Investment: Total direct investment into your company to date. Ideally it would be useful to break out personal/private vs venture funding. The intention of this ask is to highlight the "replacement cost" or "lost investment" should we allow the sector to fall to our adversaries.
11. With respect to Employees: Current employee count, broken down by full-time and part-time. This represents what we would consider a national asset: a strategic, highly-specialized and not-easily reconstructed workforce.

# APPENDIX E
## ACRONYMS & ABBREVIATIONS

AFRL – Air Force Research Lab
AFRL/RV – Air Force Research Lab Space Vehicles Directorate
AFSPC – Air Force Space Command (now USSF)
ASAT – Anti Satellite
ASDRE – Associate Director for Research and Engineering [DoD]
AST – Office of Commercial Space Transportation [FAA]
BEA – Bureau of Economic Analysis
CARES – Coronavirus Aid, Relief, and Economic Security Act
CASC – China Aerospace Science and Technology Corporation
CCL – Commerce Control List [DOC]
CFTC – Commodity Futures Trading Commission
COA – Course of Action
COVID-19 Coronavirus Disease 2019
CSRO – Chief of Space Operations Staff [USSF]
DARPA – Defense Advanced Research Projects Agency [DoD]
DFC - US International Development Finance Corporation
DISA – Defense Information Systems Agency [DoD]
DIU – Defense Innovation Unit [DoD]
DOC – Department of Commerce
DoD – Department of Defense
DOEd – Department of Education
DOS – Department of State
DOT – Department of Transportation
DPC – Domestic Policy Council [EOP]
EOP – Executive Office of the President
F2T2EA – Find, Fix, Target, Track, Engage, Assess
FAA – Federal Aviation Administration [DOT]
FBI – Federal Bureau of Investigation
FCC – Federal Communication Commission

FFRDC – Federally Funded Research and Development Center
GAO – General Accounting Office
GEO – Geostationary Earth Orbit
GEOINT – Geospatial Intelligence
ISS – International Space Station
ITAR – International Trafficking in Arms Regulation
LEO – Low Earth Orbit
MEV – Mission Extension Vehicle
NASA – National Aeronautics and Space Agency
NDRC – National Development and Reform Commission [China]
NDS – National Defense Strategy
NEC – National Economic Council [EOP]
NGA – National Geospatial Agency [DoD]
NIAC – NASA Innovative Advanced Concepts
NMS – National Military Strategy
NOAA – National Oceanic and Atmospheric Agency [DOC]
NRL – Naval Research Lab [DoD]
NRO – National Reconnaissance Organization [DoD]
NSC – National Security Council [EOP]
NSF – National Science Foundation
NSIC - National Security Investment Capital [DoD]
NSpC – National Space Council [EOP]
NSS – National Security Strategy
NSSL – National Security Space Launch
ODNI – Office of the Director of National Intelligence
OMB – Office of Management and Budget [EOP]
OSC – Office of Commercial Space [DOC]
OSD – Office of the Secretary of Defense [DoD]
OTMP– Office of Trade and Manufacturing Policy [EOP]
OSTP – Office of Science and Technology Policy [EOP]

OUSD R&E – Office of the Undersecretary of Defense for Research and Engineering
PNT – Positioning, Navigation and Timing
POTUS – President of the United States
PPP – Paycheck Protection Program
PRC – People's Republic of China
R&D – Research and Development
RDT&E – Research Development Test and Evaluation
ROTC – Reserve Officer Training Corps
RSGS – Robotic Servicing of Geosynchronous Satellites [DARPA]
S&T – Science and Technology
SBA – Small Business Administration
SBIR – Small Business Innovative Research
SDA – Space Development Agency
SIS – Space Information Services
SMC – Space and Missile Center [USSF] now SSC
WWII – World War Two

SpaceX – Space Exploration Technologies (company)
SpEC – Space Enterprise Consortium
SSC – Space Systems Command [USSF] formerly SMC
SSPIDR – Space Solar Power Incremental Development and Research [AFRL]
STEM – Science Technology Engineering and Math
STTR - Small Business Technology Transfer (STTR)
ULA – United Launch Alliance (company)
US – United States
USGS – United States Geological Service
USML – US Munitions List [DOS]
USSF – United States Space Command [DoD]
USSPACECOM – United States Space Command [DoD]
VP – Vice President of the United States

*This page left intentionally blank.*